Integrated Water Meter Management

Integrated Water Meter Management

Edited by
Francisco Arregui,
Enrique Cabrera Jr and Ricardo Cobacho

Published by   **IWA Publishing**
                 Unit 104–105, Export Building
                 1 Clove Crescent
                 London E14 2BA, UK
                 Telephone: +44 (0)20 7654 5500
                 Fax: +44 (0)20 7654 5555
                 Email: publications@iwap.co.uk
                 Web: www.iwaponline.com

First published 2006
© 2006 IWA Publishing

Disclaimer
The information provided and the opinions given in this publication are not necessarily those of IWA
or of the authors, and should not be acted upon without independent consideration and professional
advice. IWA and the authors will not accept responsibility for any loss or damage suffered by any person
acting or refraining from acting upon any material contained in this publication.

British Library Cataloguing in Publication Data
A CIP catalogue record for this book is available from the British Library

ISBN: 9781843390343 (hardback)
ISBN: 9781789065343 (paperback)
ISBN: 9781780402253 (eBook)

Contents

Preface

There were many objectives set at the beginning of this project, an initiative that may ultimately be dated back to 1999 when Francisco Arregui finished his PhD thesis on management of water meters. Many of those objectives resulted from the fact that there were not many other books on the matter and consequently many were the ways that we thought could be chosen to face the topic of water meters. This book was from the beginning conceived as both a technical reference tool and as a best practice management manual. In plain words, we always thought it would be nice to have a single book that would explain the basic principles on how meters work and that would help to make the right decisions in meter related matters.

Whether we have been able to achieve our objective it is obviously for the reader to decide. However, a great effort has gone into translating years of applied research and fieldwork into this volume. We have strived to provide not only the information that we thought was useful and relevant, but also to deliver it in a friendly and suitable manner. A great effort has been placed in including adequate photographs and figures to illustrate the concepts and ideas of this book. An effort that, despite the fact that this is a technical book, we think readers will appreciate and it will enhance their experience with the text. Another key objective for us has been to include key figures on meter performance. Most of the graphs and tables presented are the result from our own work at the laboratory and on the field. We are certain that readers with a technical background will find many of these data of great interest. Finally, we have aimed to provide up-to-date information regarding standards and technical documents. Truly a necessity in this field in which the metrology is a fundamental part of almost any process.

When this book was conceived, its main objective was to fill what we thought was a large empty space regarding literature on water meters. We still think that much remains to be said and done on the matter. Nevertheless, we hope that this work will start to fill some of that space by becoming a useful tool to water professionals around the world.

The authors

Acknowledgements

Writing the acknowledgement of a book like the one you are holding right now is always a gratifying experience. It usually means that the many hours that have been devoted to its creation are close to an end, and the result of such work is already tangible. It is then when all the people involved in this long journey, to some extent or the other, come to mind. The least we can do is to name those who helped us getting here.

This book was fully developed within the Instituto Tecnológico del Agua (ITA) and the Universidad Politécnica de Valencia (Valencia, Spain). We would like to thank our colleagues Enrique, Jordi and Vicent for their support and guidance. We would also like to show appreciation for the researchers that at different stages have been involved in this line of work at the ITA, and especially Virginia and Laura who can surely claim at least some bits of this book.

We must also acknowledge the active collaboration of several water meters manufacturers that have provided us technical support, information and great pictures and figures. In alphabetical order Actaris, Contazara, Elster and Sensus. They have made our job of describing and characterizing all devices much easier.

We would also like to thank Aguas de Valencia and Fomento Agricola Castellonense (FACSA) for the support given during these years of research.

Further mention is needed for the Spanish Ministry of Education and Science for partially funding this work through the project DPI2000-657.

Finally, we must admit that none of this could have been possible without our families and friends. Their continuous support has been crucial to bring this task to what we think is a successful end.

About the authors

Francisco Arregui – farregui@ita.upv.es

Born in 1972 in Madrid, Spain, Francisco Arregui is an Industrial Engineer with an MSc in Efficient use and Management of Water and in 1999 he obtained his PhD degree with a thesis on the management of water meters. Francisco is a senior researcher for the Instituto Tecnológico del Agua, and his research is focused in water meter technology and management. He has been involved in several international projects on the topic, acting as an external consultant for water companies. Francisco Arregui is member of the IWA taskforce on water loss and is an active participant in the Spanish standardization committee in charge of metrology.

Enrique Cabrera Jr. – qcabrera@ita.upv.es

Born in Vila-real, Spain, in 1972, Enrique Cabrera graduated in Industrial Engineering. MSc in efficient use and management of water, obtained his doctor degree with the thesis "Design of a performance assessment system for urban water utilities". Enrique is associate professor of fluid mechanics at the Polytechnic University of Valencia. He is secretary of the IWA Specialist Group on Efficient Operation and Management of Urban Water Systems and leader of its Benchmarking Taskforce. Enrique is actively participating in the ISO Technical Committee 224, dealing with "Service activities relating to drinking water supply and wastewater".

Ricardo Cobacho – rcobacho@ita.upv.es

Born in 1972 in Melilla, Spain, after obtaining his degree in Industrial Engineering, Ricardo completed MSc and PhD degrees focused on the efficient use of water and demand management topics. Since then, he has been working on topics related to urban water distribution and its efficient use and management. Ricardo is an associate professor of fluid mechanics at the Polytechnic University of Valencia. He is an active member of the IWA taskforce on Efficient Use and Management of Water.

The authors can be contacted at:
Instituto Tecnológico del Agua, Universidad Politécnica de Valencia
Apdo. 22012, 46071 Valencia – Spain
Tel: +34 96 387 98 98 – www.ita.upv.es

1
Introduction

1.1. INTRODUCTION

The absence of specific literature about water meters is quite striking, both from a technical and from a managerial point of view. Despite the fact that water meters as we know them have been in use for over 100 years, and that measuring the consumption of water has been a necessity for thousands, the number of comprehensive references about the topic is quite scarce.

However, reality shows that this apparent lack of interest in the matter does not correspond with the needs and expectations of technical and managerial staff in charge of water distribution systems worldwide. In the authors' experience, the continuous demand for technical workshops, reports and projects shows that the water industry worldwide needs to learn many more things about water meters in a subject that remains relatively unknown to most people in the business.

This is even more surprising considering that meters are one of the cornerstones of the business of supplying water in most parts of the world. It is a well-known fact that in the first page of the manual of most private operators, installing or reviewing water meters is almost the first action to undertake when taking care of a new utility. Not a cause of astonishment, considering that meters are the sole instrument available to provide fair and documented billing. Furthermore, meters provide data that may be fundamental for the technical management of a utility, allowing for rigorous water balances, characterizing demand and identifying anomalous behaviours and failures in the system.

From a managerial point of view, water meters may even be more important. Water price has been increasing for the past decades around the globe, in an increasing attempt to reflect the real costs of abstracting, treating and distributing drinking water to the end users. The more water costs, the more costly that leaks, illegal consumptions and any other kind of losses become. Furthermore, the more water costs, the higher is the impact on revenue of all meters installed in a utility.

Consider a water utility which is facing the replacement of an important part of the water meters which are installed in the system. There are currently many options in the market in terms of technologies, manufacturers and quality. Even with meters fully conforming to the latest standards, the range of options available to the decision makers is undoubtedly vast. However, experience shows that this sort of decision is often made with little rigorous information at hand. Managers are often inclined to decisions driven by "tradition" (the cheapest option, a certain model, a technology or a manufacturer who has delivered what is considered to be a good performance) or simply by the very limited information available or other external factors.

However, this decision which is sometimes overlooked, other times left to be made by the wrong person, and most of the times not meditated or prepared enough may be one of the single decisions to have most impact on the business by the company. The selection of a certain meter model or size in favour of another may have a repercussion in the total registered volume of several percentage points (e.g. 3%). A percentage which, depending on the tariffs, will be directly reflected or even increased into the billed amount, and consequently on the revenue of the utility.

Choosing the right meter, or even the time to replace it is not an easy decision to make. Many factors need to be taken into account, and some of them are seldom known well enough to be certain of making the right choice. However, this book is aimed to provide both the information and the methodology that will lead decision makers to look into the right directions when faced with such choice.

1.2.   AN INTERDISCIPLINARY MATTER

As a reader of this book you will be faced with contents of very different nature. Meter management is truly an interdisciplinary matter and notions of topics in different fields are required to fully understand all the factors involved in the process.

Engineers will feel comfortable with the technical items. Meter construction and technology is better understood with some fluid mechanics knowledge, and those with experience in mechanical engineering will easily follow some of the issues involved in the construction and development of water meters.

Many of the techniques developed for optimum meter selection and the replacement frequency are closely linked to statistical concepts. Water meter management requires dealing constantly with uncertainties and their assessment and propagation. The necessary data on water demand is usually gathered from sampling or data mining from the commercial databases. In any of those cases, statistical preparation (for instance sample selection) and processing of the data is required.

Finally, it is difficult to assess the advantages and disadvantages of one model vs. another one without some basic economic knowledge. The same applies to the renovation frequency of the meters. Obviously, utility managers should find no trouble whatsoever in following the economic assessment of the different available options presented in this book.

In any case, far from trying to discourage the reader, we think that this variety of topics only adds further interest to the problems at stake in this book. The need to explore areas which may not be familiar to some readers will later pay off by providing an improved overall view of the different aspects involved in water distribution.

1.3. STRUCTURE OF THIS BOOK

The present book has three differentiated parts, although they are closely related and most readers will not get the full benefit from the information contained in some of them until they have read the rest. However, the chapter structure has been designed to facilitate a practical use of this book, and the different parts while bearing a close relationship are mostly self-contained. Consequently, a full journey through this book will provide a comprehensive understanding of all key issues regarding the integrated management of water meters, from selection to renewal. This will not prevent the fact that the reader is likely to return later to individual chapters or parts of this book for specific reference.

The first part of the work, which spans Chapters 2–4, describes the different devices subject of this book. Chapter 2 deals with the available technologies for water meters, their construction, characteristics and presents the advantages and disadvantages of each one of these technologies. Analogously, Chapter 3 describes the different alternatives that can be found for flow meters. Finally, Chapter 4 deals with the new technological developments that have enabled to read meters remotely and systematically, with all the associated advantages.

The second part of this book, comprised of Chapters 5–7, deals with the management of a population of meters from the different perspectives involved in the process. Chapter 5 presents the management of water meters from an integrated approach, rather than considering the different processes involved as independent actions. All factors, from the initial selection of the meter, through the quality control at reception and their maintenance influence the overall performance of the meters population and should be taken into account. In order to provide further detail on some of these aspects, Chapter 6 deals with the appropriate assessment of the overall meters' accuracy, while Chapter 7 presents the problem of maintenance and renewal of meters in a utility from an economic perspective in order to maximize the return on the investment.

The third part of this book presents practical information to perform some of the processes involved in the management of meters. Chapter 8 describes the principles and procedures for meter testing. Chapter 9 is aimed to facilitate an adequate quality control at the reception of new meter batches. And finally, Chapter 10 presents a detailed overview of the main standards and technical guidelines that apply to water meters in different parts of the world.

2
Water meters

2.1. SINGLE JET METERS

2.1.1. Operating principles

Single jet meters may be considered under the category of velocity meters. The single jet technology is the most widely used in many countries around the world to meter domestic users due to its low cost and great reliability.

The operating principles are based on the tangential incidence of a *single* jet over a radial-vaned impeller placed inside the body of the meter. The rotation velocity of the impeller is proportional to the impact velocity of the water, or in other words, to the circulating flowrate. Any modification in the existing relationship between the circulating flowrate and the impeller's rotation velocity will alter the error curve, and consequently, the metering error.

2.1.2. Constructive characteristics

Figure 2.3 shows the different parts in a single jet meter. The *impeller* can be found inside the meter *housing*, which is usually made of brass or bronze. The internal housing dimensions are critical for the accuracy of the device, and the strict tolerances of the manufacturing process lead to higher prices. For this reason, single jet meters are mostly used in diameters ranging from 7 to 20 mm, where the savings in the amount of material used with respect to multijet meters justifies a more expensive manufacturing process. However, in medium diameters (between 25 and 40 mm) the differences

FIGURE 2.1.   OPERATING PRINCIPLE OF A SINGLE JET METER

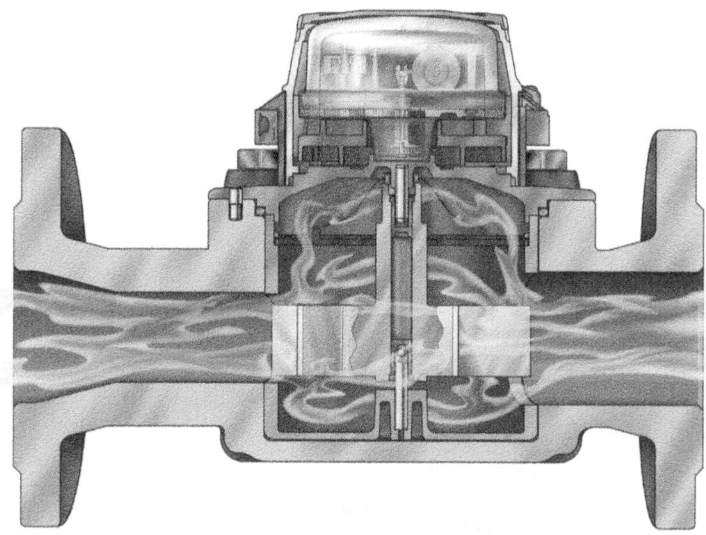

FIGURE 2.2.   LONGITUDINAL VIEW OF A SINGLE JET METER (COURTESY OF SENSUS)

are not as significant, and multijet meters turn out to be a more economic option when comparing meters of similar metrological performances.

The base of the measuring chamber, where the impeller is placed, may be flat or with guides. Their function is to generate turbulent flow at low flowrates so the error curve remains within the standardized limits for a wider range of flowrates. However, the presence of guides also creates a hammering effect that in the medium term may deteriorate the supporting points of the impeller, which may derive in a premature wear of the instrument.

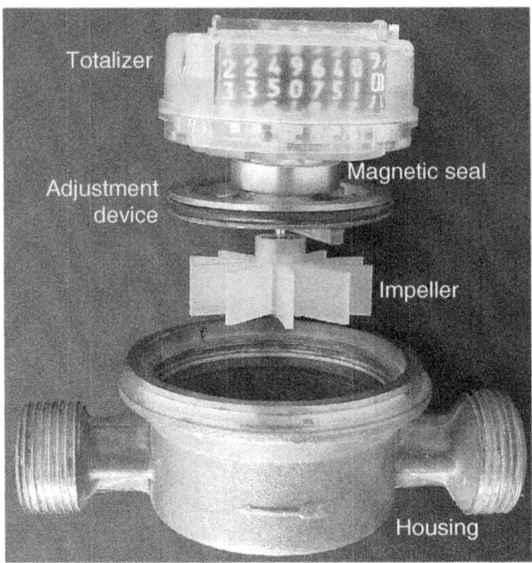

FIGURE 2.3. PARTS OF A SINGLE JET METER

FIGURE 2.4. PLASTIC CHAMBER IN A SINGLE JET METER

Some manufacturers are currently using a different meter arrangement which derives from multijet meters. The metal housing actually holds a plastic module where the impeller is placed, as shown in Figure 2.4. In this way, the manufacturing costs of the metal housing are considerably reduced, for the tolerances do not need to be as tight. Since the module acting as measuring chamber, is made out of plastic, its manufacturing tolerances can be set very low. Additionally, it is also important to note that there

FIGURE 2.5.   DIFFERENT TYPES OF SINGLE JET METER IMPELLERS

are specific plastics that can be used to avoid calcium sediments and all its related side effects.

The pivot of the impeller may be found in the meter housing or the housing cover (Figure 2.3). The *impeller* is in any case, the element that transforms the linear velocity of the flow into circular motion. It is usually manufactured in plastics with a density somewhat lower than water. The aim is that the impeller will float and only contact the rotation pivot at one point, reducing the friction losses. The tip of the rotation pivot usually features a synthetic ruby or sapphire, or any other element that is wear resistant. It is also quite frequent that manufacturers search for a combination of materials for the impeller and the bearings that are both resistant to wear and minimize friction.

Figure 2.5 shows several types of impellers. The design has changed quite a lot from the early designs (left). In the latest models, a magnet is fixed to the top of the impeller. This magnet acts as a coupler with the follower magnet placed in the totalizer, allowing for a transmission of the motion without the need of physical contact between the parts.

Many meters also incorporate magnetic seals to avoid fraud. The seals are ring shaped and placed around the two magnets (placed at the impeller and the totalizer). The presence of this seal avoids the blocking of the gear train by means of an external magnetic field.

In other models, the transmission of the motion between the impeller and the totalizer is purely mechanical. The coupling is done by gears instead of magnets. This solution is usually adopted in Class C meters, with wet or semi-wet dial meters, and in the oldest models.

As far as the size of the impeller is concerned, a larger impeller delivers higher torque for a given flow velocity, and as a consequence the meter will start at a lower flowrate.

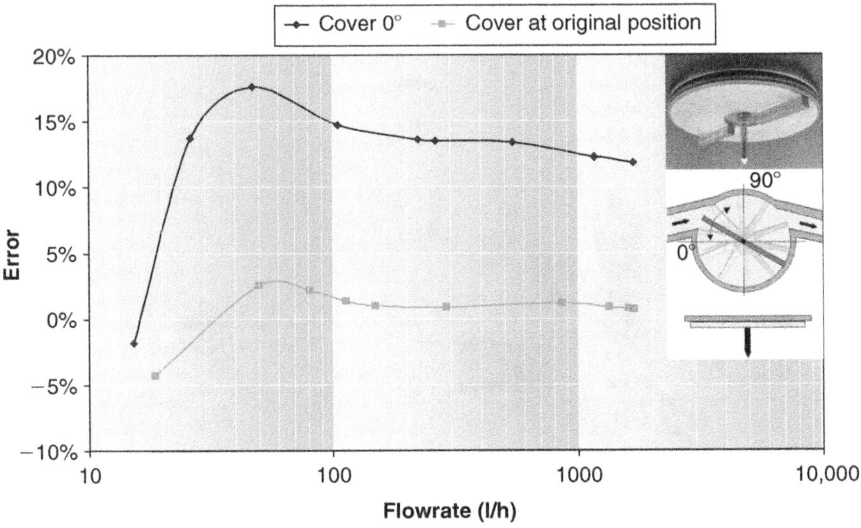

FIGURE 2.6. DISPLACEMENT OF THE ERROR CURVE BY USE OF THE ADJUSTING DEVICE

However, the additional inertia created by a larger diameter of the impeller must also be taken into account when determining the effects of the diameter on the starting flowrate. But, a large impeller implies higher cost of the meter due to the material used to manufacture the housing.

The *cover of the measuring chamber* is often used as an adjusting device. It usually presents one or two guides that allow adjusting the error curve of the meter during the calibration process. The regulation is achieved by turning the cover, which plays the role of a hydraulic brake. Following this method, the whole error curve can be displaced about a ±15% in the whole range of flowrates.

As an example, Figure 2.6 shows the effect of turning the cover of the measuring chamber on the error curve of a certain single jet meter. In the tested meter, when the cover is placed at 0° (instead of the original factory position) the error curve is displaced to the zone of positive errors, overregistering in approximately +13%.

Setting the adjusting device at 0° reduces for a given flowrate the drag torque of the impeller, and consequently the impeller rotates faster. By setting the cover at 90°, perpendicular to the flow direction, the drag torque is increased and the impeller rotates slower with the same flowrate.

In any case, this type of adjusting device is designed to be used only at the manufacturing stage, and cannot be manipulated without breaking the calibration seals, once it has been installed.

Other models of single jet meters use a by-pass circuit in order to adjust the error curve. Once the water enters the meter housing, it can follow two different paths to reach the exit. One of them runs through the measuring chamber where the impeller is located, while another one of much less capacity runs parallel to it. The flowrate derived to the parallel circuit, which does not impinge on the impeller, may be regulated by adjusting a screw (Figure 2.7). The larger the derived flowrate is, the smaller is the energy

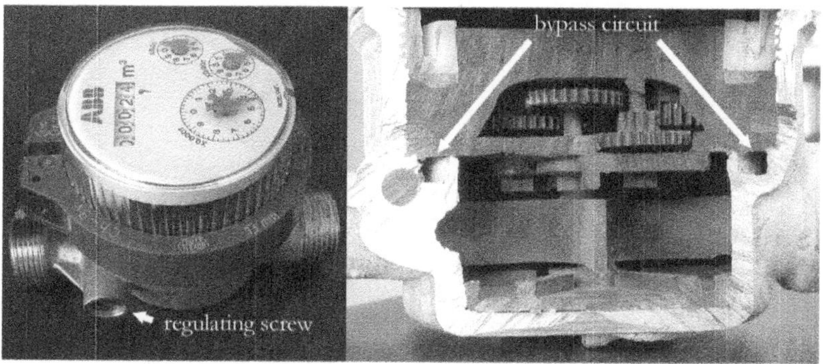

FIGURE 2.7. SINGLE JET METER REGULATED BY A BY-PASS CIRCUIT

TABLE 2.1. TYPICAL DIMENSIONS AND METROLOGICAL CHARACTERISTICS OF
SINGLE JET METERS

Diameter (mm)	Qn (m³/h)	Length (mm)	Metrological classes
7	0.6	115	B
13	1.5	100, 115, 190	B, C
15	1.5	100, 115, 190	B, C
20	2.5	115, 190	B, C
25	3.5	260	C
32	6	260	C
40	10	300	C
50	15	300	C
65	20	300	C
80	30	350	C
100	50	350	C

transferred to the impeller, which consequently rotates slower. On the other hand, a higher throttle in the by-pass displaces the error curve towards the zone of positive errors.

2.1.3. Metrological characteristics and dimensions

Single jet meters are manufactured in a wide range of diameters ranging from 7 to 100 mm, both in Classes B and C. The most popular ones in the range are the small meters (13, 15 and 20 mm, Classes B and C).

Table 2.1 shows the most common characteristics that can be found for the different diameters, including the nominal flowrate, the metrological class and the typical lengths available in the market. It is important to point out that for a given diameter, the length is not unique, and this may turn out to be quite an inconvenience. As a matter of fact, many utilities find this out when replacing certain meters, which is the reason why an updated database should be kept with the meter data.

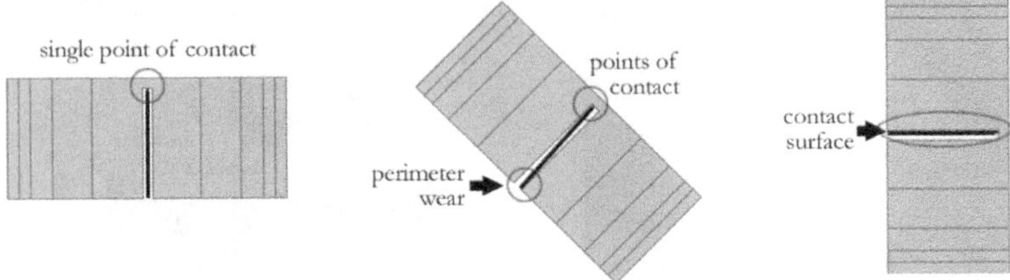

FIGURE 2.8. IMPELLER CONTACT WITH THE ROTATION AXIS IN A TYPICAL SINGLE JET METER

2.1.4. Metrology of single jet meters: Orientation

Single jet meters are usually designed to operate horizontally, which is equivalent to say with their rotating axis perfectly vertical. This position, depending on the construction characteristics of the meter, is the only one that allows the impeller to rest only on the pivotal point, as shown in Figure 2.8. This obviously minimizes friction and measuring errors at low flowrates are not affected, for this is the design position of the meter. If properly installed, the wear of the rotation axis and its tip is low, for they are manufactured with materials of an exceptional hardness and resistance to wear.

When the meter is inclined, the impeller makes contact in different areas that have not been designed for such contact, increasing friction. Consequently, for a given flowrate, the impeller will rotate slower due to an increased friction torque, and the registered volume will be smaller than the actual volume passed through the meter. This phenomenon is only noticeable at low flowrates, where the energy transferred from the water to the impeller is small, and any slight modification of the forces balance affects the rotation velocity of the impeller.

The worst situation appears when the impeller rests on the side of the rotation axis. In this case, depending on the design, it is not possible to talk about contact points anymore, but a contact surface. Friction in this position increases notably and so do the errors at low flowrates. Generally speaking, a Class B meter will turn into Class A when installed in a non-horizontal position.

Figure 2.9 shows how the orientation of a meter (1.5 m^3/h nominal flowrate, Class B) during installation affects the error curve. For flowrates above 100 l/h, the degree of inclination of the meter has practically no consequences on the metering error. However, the meter is much more sensitive to orientation at low flowrates (i.e. 60 or 30 l/h).

Another key factor which justifies a correct installation of a meter is the starting flowrate. As a matter of fact, an incorrect orientation of the meter will not only affect the accuracy at low flowrates, but it will also increase the value of the flowrate that starts the rotation of the impeller. A large value for this parameter significantly reduces the chances of detecting leaks in the user's facilities, leaks that in any case will have to be assumed by the water service provider.

In order to determine the influence of the orientation on the starting flowrate, several experiments were carried out on used Class B meters (6–9 years in age). The

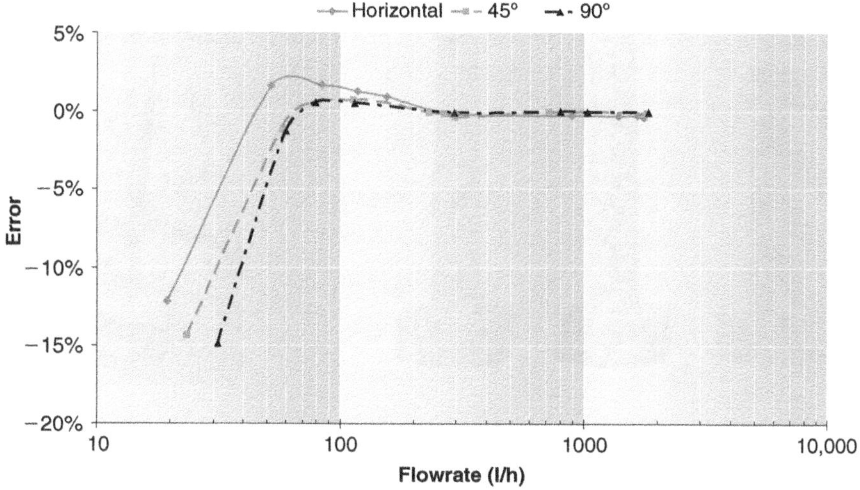

FIGURE 2.9. SINGLE JET METER'S ERROR CURVE AS A FUNCTION OF ORIENTATION

results of the test were clarifying, with an average increase of 10 l/h in the starting flowrate (from 24 to 34 l/h) with a 45° inclination.

However, it is important to take into account the fact that although most single jet meters are affected by an incorrect orientation, the effect depends on the actual model (some models do a better job at maintaining their accuracy in an incorrect orientation). This fact is especially relevant in utilities where a significant number of meters are expected to be installed with an incorrect orientation. In such cases, the most appropriate models in the market should be selected.

Some designs achieve a better behaviour by holding up the impeller in two points that restrict vertical movement while allowing its rotation. This design would reduce, in theory, the inconveniences derived from incorrect impeller supports that appear in inclined meters.

As a matter of fact, some meter models are approved to work both in horizontal and vertical positions while maintaining their metrological characteristics (European Directive 75/33/EEC). In these cases the meter will be conveniently marked, indicating its class for all positions or otherwise indicating the metrological performance for each orientation of the meter.

In addition to an increase for low flowrates, an incorrect installation may lead to an excessive wear of the impeller parts in contact with the pivot (Figure 2.10). This will originate in the medium term an increase in friction and metering errors, thus reinforcing the recommendation of a proper orientation in the installation of meters.

2.1.5. Metrology of single jet meters: Velocity profile

As velocity meters, the error curve of single jet water meters may be affected by changes in the velocity profile. This means that the presence of a valve or a change in direction of the pipe near the entry point of the meter might change its accuracy.

FIGURE 2.10.   INCREASING WEAR IN THE IMPELLER BEARING DUE TO AN IMPROPER ORIENTATION OF THE METER

FIGURE 2.11.   OBSTRUCTED METER STRAINERS

In practice, metrology is hardly ever significantly affected by this phenomenon. The design of the meters can in part account for it:

- The nozzle that leads to the impeller chamber often has a convergent shape. The flow is accelerated at this point, and the velocity profile regularized before it reaches the metering chamber.
- All single jet meters distort to a certain extent the velocity profile (distortion caused by changes in the direction or cross section). These flow alterations tend to minimize the effect on the error curve of changes in the velocity profile due to other factors.

Another factor that can lead to changes in the error curve of the meter is the obstruction (partial or total) of the entry strainer, which is actually a quite common situation (Figure 2.11). As mentioned before, the velocity profile is regularized at the entry nozzle, and the effect of these irregularities is in most cases greatly minimized. However, this circumstance may be significant in short single jet meters (100 mm in length) with a much shorter entry nozzle (Figure 2.12).

A special situation takes place when the obstruction is provoked by the rubber gasket in the meter's connection, due to its being too tight. In these cases, depending on the type of rubber, diameter of the pipe and meter design, some significant over-registering can take place. In extreme cases, when the reduction in the cross section caused by the gasket is considerable, and the jet is concentrated in the centre, error values can rise up to $+10\%$.

FIGURE 2.12.   ENTRY NOZZLE LENGTH FOR METERS OF 115 MM (LEFT) AND 100 MM

TABLE 2.2.   INFLUENCE OF UPSTREAM PERTURBATIONS IN METERING ERROR

Test	No valve (%)	Fully open valve (%)	Halve open valve (%)	Partially blocked strainer (20%) (%)
1	−0.2	0.7	1.5	0.5
2	−0.3	−0.6	−0.9	
3	−0.2	0	1.2	
4	0.9	1.5	1.4	0.9
5	0.8	1.2	1.3	0.8
6	0.9	1.6	1.3	0.7

Table 2.2 shows the results of a series of tests carried out for several 115 mm meters at an approximate flowrate of 1350 l/h. The error figures compare a situation with no upstream perturbation, with:

- Fully opened valve upstream.
- 50% opened valve upstream.
- Partially obstructed strainer (20% of the cross section).

Table 2.2 shows that for all meters, the differences in the metering error are hardly significant and always below 1%. Consequently, it can be stated for most single meters that this sort of distortion does not affect the final result of the measure.

2.1.6.   Metrology of single jet meters: Aging

The rate of decay of the error curve of a meter with age is a crucial factor for optimizing meter management and determining the optimum replacement period.

The first noticeable effects of aging take place at low flowrates. Any increase in resistance to the rotation due to mechanical wear will have a greater effect when the energy transferred to the impeller is lowest, and consequently at low flowrates.

As a matter of fact, the starting flowrate may increase considerably while the error at medium flowrates (close to the nominal value) remains the same. Figure 2.13 shows this same situation for a Class B, 15 mm calibre meter. The starting flowrate exceeds

60 l/h (six times higher than the usual value for a new meter with those characteristics). For a flowrate of 200 l/h, the error is still high (-19%), but it keeps decreasing for higher values and at 750 l/h is only -3%.

The error at medium and high flowrates may be both negative and positive. In the case of a positive error (higher registered volume than the actual one) it can be caused by the reduction in the free space between the impeller and the housing due to the accumulation of sediments in the impeller chamber. As a rule of thumb, the positive errors are not usually too high and seldom exceed 10%. However, there are cases like the one shown in Figure 2.13 (below), where the metering error was $+25\%$ for almost all the measuring range.

In the medium term, the limescale build up, or calcium deposits (Figure 2.14) can even temporarily block the impeller until a water transient generates enough momentum for its release. This effect would show up in the billing databases as a gradual decrease in the registered volumes, as the impeller gets blocked more and more often

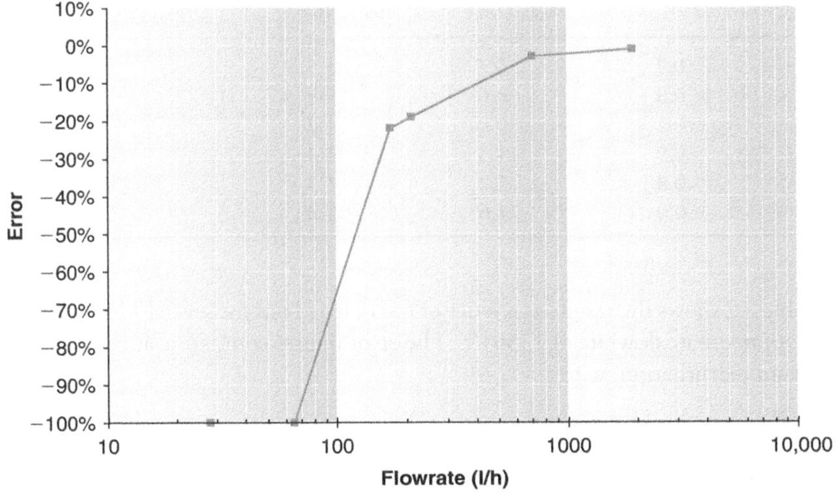

FIGURE 2.13. ERROR CURVE FOR A 9-YEAR OLD CLASS B, 15 MM METER

FIGURE 2.14. LIMESCALE BUILD-UP IN A SINGLE JET WATER METER HOUSING

when deposits grow. At a final stage the impeller would be blocked permanently. In those utilities with hard water, in which limescale build-up is a problem, the selection of meters according to their material and surface finishing is a key factor in guaranteeing a minimum period of life for the meters.

Another key constructive factor influencing the downgrade of the error curve is the regulation system of the meter. Those meters adjusted by means of a hydraulic brake (Figure 2.6) are rarely affected by positive errors. However, those using a by-pass circuit to adjust the error curve (Figure 2.7) run the risk of getting the by-pass clogged, generating positive errors at medium and high flows (the volume that should be going through the by-pass is registered by the impeller). This fact has been experimentally checked at the laboratory, although only moderate errors, close to $+5\%$, should be expected. For low flows, the additional volume going through the measuring chamber does not compensate the negative error due to the increment of the drag torque caused by the mechanical wear of the meter, and the resulting error is still negative.

Since the metering errors are not constant for all values of the flowrate, it is not possible to define a generic pattern for the accuracy decay of a meter, for it depends on the flow that is being registered. A meter with a poor performance for low flows may be accurate enough in the case of higher flows being registered. Additionally, errors may compensate themselves, and the underregistration that takes place at low flows due to mechanical wear can be compensated by positive errors at medium and high flowrates.

2.1.7.   Metrology of single jet meters: Reverse flow

In systems that are not permanently pressurized and service interruptions take place, reverse flows through the meters are likely to occur. These reverse flows can be prevented by the use of non-return valves. In this situation, meters should keep functioning, and even subtracting negative volumes from previously registered ones (new ISO standards require that if they register this reverse flow, they do it with the same level of accuracy as forward flows). Table 2.3 shows the error of five single jet meters – same model – at four different flowrates, in both directions.

It should be mentioned that different meter models could provide different results, depending on their design and diameters of the entry and exit nozzles. The exit nozzle

TABLE 2.3.   METERING ERRORS FOR A 15 MM METER (FORWARD AND REVERSE FLOWS)

Meter	2000 l/h		655 l/h		200 l/h		55 l/h	
	Reverse (%)	Forward (%)	Reverse (%)	Forward (%)	Reverse (%)	Forward (%)	Reverse (%)	Forward (%)
1	−44.4	1.6	−42.5	0.4	−43.0	0.3	−51.7	−9.9
2	−45.3	2.5	−45.3	1.5	−48.4	1.9	−43.6	−5.8
3	−47.9	−0.4	−47.5	−1.9	−48.4	−1.8	−47.5	−8.5
4	−45.2	0.2	−45.6	−1.4	−50.7	−1.2	−46.5	−0.4
5	−47.9	0.5	−48.3	−0.6	−49.1	−0.2	−43.0	−7.2
Average	−46.1	0.9	−45.8	−0.4	−47.9	−0.2	−46.4	−6.4

is usually larger, producing smaller velocity values for equal flows when these take place in the reverse direction. Consequently the impeller rotates slower and registers smaller values.

These are unfair conditions for the user, since some of the registered volume is never consumed as it returns back to the system through the meter. According to the values in Table 2.3, approximately half of this "two-way" volume is registered and consequently billed to the user.

2.1.8. Pulse emitters

Despite the fact that the use of pulse emitters in meters of small diameter is not very common (due among others to the low reliability of the most common models) it is convenient to take into consideration their possible effects in the registration of volumes.

It has already been mentioned that the error of a meter at low flows depends to a great extent of the resistance torque present at the impeller. However, another key factor is the inertia of the system (impeller and gears). The greater the inertia, the longer it will take for the impeller to rotate at the right speed and register the flow adequately. This is the reason why adding a REED pulse emitter (consistent of a magnet placed on the gears) will increase the inertia of the system, and the value of the starting flowrate.

This effect increases as a factor of the relative volume per pulse. If the magnet is placed on a wheel that will turn once for every litre registered, the inertia of the system will be greater than when placed on a wheel that will turn once for every cubic meter. In this second case, the influence of the magnet may be negligible, while in the first case it could increase the starting flowrate in a single jet Class B meter from 10 to 20 l/h.

In any case, the influence that this sort of emitters can have in the error curve at flowrates of 60 l/h and higher is negligible, regardless of the volume associated to each pulse (Arregui, 1999).

2.1.9. Advantages and disadvantages of single jet meters

Advantages	Disadvantages
• Resistant to suspended solids. Adequate for hard waters or those with a number of suspended particles.	• The impeller, its bearing and the mechanical parts are likely to wear in non-ideal conditions.
• Great variety and availability. Easy to find the right model for the needs.	• They are generally affected by the installation orientation.
• Reliable technology, used for decades.	• To date, no Class D ISO 4064:1993 model available.
• Small and can be installed in tight spaces.	
• 13, 15 and 20 mm meters are probably the cheapest that can be found in the market among all technologies.	• In shorter meters, clogging of the strainer or a gasket fitted too tight may lead to errors.
• Class C single jet meters are a good substitute for combination meters for large users with a wide range of flows	• By-pass regulated meters may produce positive errors at higher flows. Conflictive with users.

(Continued)

Advantages	Disadvantages
(schools, industries, swimming pools, ...). • Relatively low sensitivity to the velocity profile.	• Starting flowrates are not low enough to detect leakage at the user's side, although latest models show improvement. Depending on construction, starting flowrates can increase dramatically with time.

2.2. MULTIPLE JET METERS

2.2.1. Operating principles

Multiple jet (multijet) meters can also be included in the group of velocity meters. They are mainly for domestic consumers, and sometimes in irrigation networks. Their operating principle is quite similar to the one featured in single jet meters. The difference lies in the fact that in multijet meters the water impacts the impeller in multiple points around its perimeter instead of doing it in a single point.

This method achieves a more balanced operation of the impeller, and in theory a greater durability of the meter. Additionally, multijet meters are supposed to behave better at low flowrates and have lower starting flowrates.

Just as with single jet meters, the impeller's angular velocity depends on the velocity of the water jets. Consequently, any modification affecting the flow or the entry velocity of water in the measuring chamber will alter the error curve.

2.2.2. Constructive characteristics

Figure 2.15 shows the different parts of a multiple jet meter. The measuring chamber can be found inside the meter housing. However, contrary to what happens in single jet meters, the dimensions of the housing have no effect on the accuracy of the meter, allowing for more flexible tolerances in the manufacturing process.

On the other hand, comparing meters of equal diameter, multijet meters have a larger housing, and therefore need more materials in their construction. For instance, a typical multijet meter for a nominal flowrate of 1.5 m³/h (15 mm diameter) weights 2 kg and measures 190 mm in length, compared to the 115 mm of a single jet one. However, some especial designs (concentric meters) allow separating the connection housing from the metering mechanism, which in some cases is even replaceable (metering insert).

For small diameters (up to 20 mm) the savings in materials are usually more important than the higher production costs, making multiple jet meters more expensive than single jet meters. However, for larger diameters the terms get inverted and multijet meters are usually cheaper to manufacture.

In multiple jet meters, the adjustment of the error curve is carried out by means of a *regulating screw* which controls the amount of flow circulating through a by-pass circuit. The water can consequently go through the meter following two different

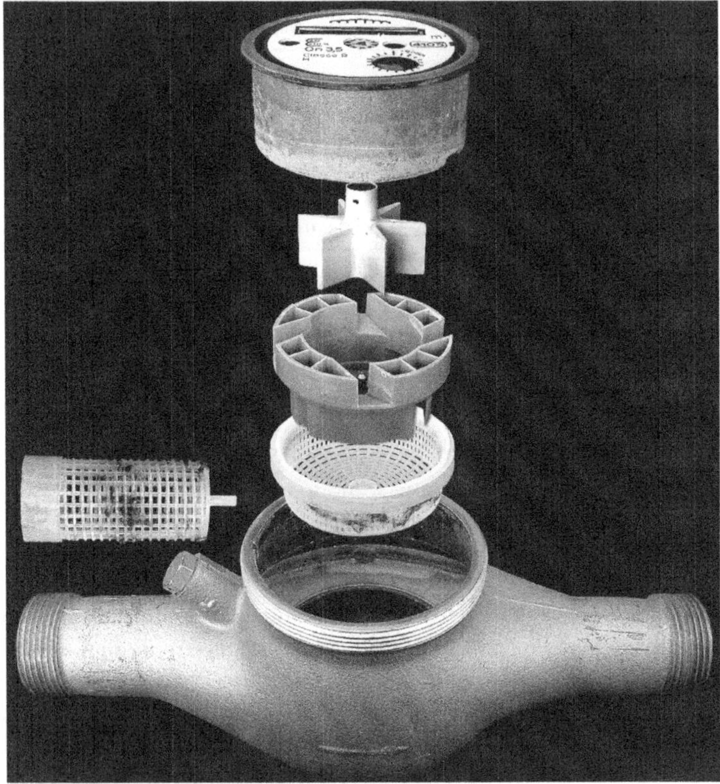

FIGURE 2.15. PARTS OF A MULTIJET METER

paths, one through the impeller chamber and the other through the by-pass, which is regulated by the screw. The higher the by-pass flowrate, the smaller the volume that gets registered by the meter while the error curve is displaced towards negative errors (underregistration). When the by-pass is closed using the screw, the flowrate through the measuring chamber is increased and the error curve is displaced towards positive values. Figure 2.16 shows the operation of the regulation of a multijet meter.

The orifices placed at the entry and exit points to the impeller chamber are oriented in different directions. This particular orientation is designed to minimize the friction losses in the meter. The impact velocity of water on the impeller only depends on the flowrate and the dimensions of the entry orifices, and as mentioned before, the geometry of the casing does not play a significant role.

Quite often, multiple jet meters present a long *strainer*, and in some cases, another one covering the openings to the metering chamber. Obviously, a clogged entry strainer will not have any effect on the metrology of the meter. However, the obstruction of the strainer located near the metering chamber could lead to an increase in the impact velocity on the impeller for the other openings. This would result in a faster rotation of the impeller for a given flowrate and in undesirable metering errors.

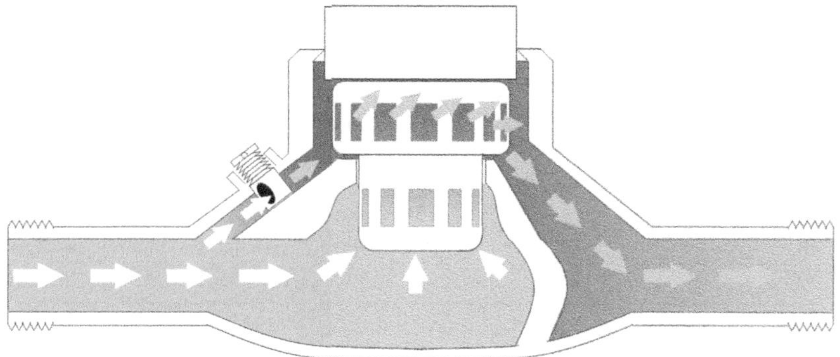

FIGURE 2.16. WATER FLOW INSIDE A MULTIPLE JET METER

This is precisely one of the inconveniences of multiple jet meters. It is not rare to observe the errors for medium and large flows drifting to positive values (overregistering) which considerably affect the quality of the measure. The obstruction of the by-pass circuit and the chamber entry strainer are the main causes for this phenomenon.

The coupling between the impeller and the totalizer can be done, as in single jet meters, both mechanically and magnetically. A mechanical coupling usually corresponds with older or Class C models, while magnetic coupling is more common and is used in Class B meters. The constructive principles are equivalent to those found in single jet meters, and the same considerations described for them apply.

2.2.3. Metrological characteristics and dimensions

Multiple jet meters are usually manufactured in diameters ranging from 15 to 50 mm. For diameters above this range, the flowrates that they may stand are small in comparison with other types of meters.

They can be found in Classes B and C, although the latter usually only corresponds to the smaller diameters. The use of multijet meters is widely extended in Latin America and some European countries. The American Water Works Association currently has a specific standard for this kind of meters.

Tables 2.4 and 2.5 show the most common characteristics for different diameters, including the metrological class, the nominal flowrate and typical lengths that are currently available in the market.

2.2.4. Metrology of multiple jet meters: Orientation

Multiple jet meters are usually designed to operate in a perfectly horizontal position. This set up reduces considerably the wear of the mobile parts. All comments made for single jet meters on this issue are also valid for multiple jet meters.

The orientation of a multiple jet meter is only relevant for low flow errors. This leads to an increase in metering errors due to leaks in the users' facilities with the consequent economical loss.

TABLE 2.4. SIZE AND CLASSES OF TYPICAL MULTIPLE JET METERS

Diameter	Qn (m³/h)	Length (mm)	Metrological classes
15	1.5	190	B, C
20	2.5	190	B, C
25	3.5	260	B
30	6	260	B
40	10	300	B
50	15	300	B

TABLE 2.5. GENERAL CHARACTERISTICS OF CLASSES B AND C MULTIPLE JET METERS

Metrological class	B	C	B	C	B	B	B	B
Diameter (mm)	15	15	20	20	25	30	40	50
Nominal flowrate (m³/h)	1.5	1.5	2.5	2.5	3.5	5	10	15
Maximum flowrate (m³/h)	3	3	5	5	7	10	20	30
Transitional flowrate (±2%) (l/h)	120	22.5	200	37.5	280	400	800	3000
Minimum flowrate (±5%) (l/h)	30	15	50	25	70	100	200	450
Starting flowrate (l/h)	7	7	12	12	25	25	35	45
Pressure loss at maximum flowrate (bar)	0.9	0.9	0.9	0.9	0.8	0.9	0.9	0.9

TABLE 2.6. AVERAGE ERROR FOR ½″ MULTIPLE JET METERS

	Starting flow (l/h)	30 l/h	60 l/h	120 l/h	500 l/h	1500 l/h	3000 l/h
Used strainer (%)	23.01	−20.8	−0.2	2.6	3.0	3.8	3.0
No strainer (%)	25.76	−17.0	−3.2	2.0	1.4	2.8	2.7
Difference (%)	2.75	3.8	−3.0	−0.6	−1.6	−1.0	−0.3

2.2.5. Metrology of multiple jet meters: Velocity profile

The velocity profile at the entry of the distribution chamber is influenced by the changes in direction taking place inside the meter's housing. As a consequence, perturbations upstream of the meter have no influence on its accuracy, and multijet meters do not precise straight sections of pipe preceding the meter to ensure an accurate measure of the flows.

In 2000, research by Arregui showed, after testing over 300 multijet meters, the influence on the starting flowrate and the error curve of dirt accumulation at the strainer placed at the inlet of the meter. First, the meters were tested with their original, and later, all the strainers were removed and the meters tested again. The results for these tests are shown in Table 2.6.

Table 2.6 shows almost no differences, and those that can be found could be attributed to other factors such as the uncertainty in reading the meters, given the low

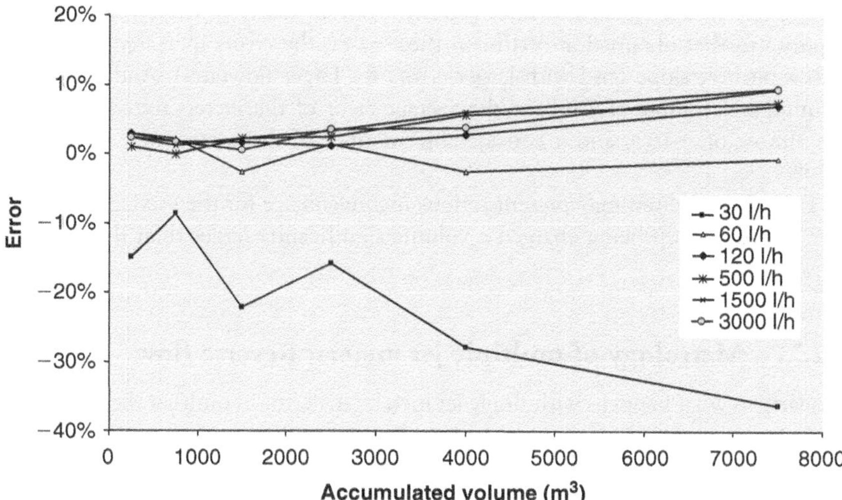

FIGURE 2.17.   MULTIJET ERROR (NOMINAL FLOW 1.5 M³/H) ON ACCUMULATED VOLUME

resolution of the totalizer. The importance of this uncertainty during the tests increases for low flows (these tests are usually carried out for small volumes for time reasons, and the minimum resolution in this case – 0.2 l – represented 2% of the total measured volume – 10 l).

The conclusion to these tests is clear. Unless the meter strainer is totally clogged, the accumulation of dirt in the entrance strainer has no influence on the error curve of a multiple jet meter.

On the other hand, the obstruction of one or several orifices at the entry of the impeller chamber increase the impact velocity significantly altering the error curve, and so does the obstruction of the by-pass circuit. Such circumstances are similar in nature to what happens to single jet meters when the cross sections of the entry nozzle or between the impeller and the housing are reduced.

2.2.6.   Metrology of multiple jet meters: Aging

The error curve of multijet meters can be displaced sometimes to the positive errors zone for medium and high flowrates (due to the partial obstruction of the by-pass circuit). In the series of tests mentioned before, it was learnt that, in average, those meters with a larger registered volume through their lives, the error at medium and high flowrates were larger. A summary of the obtained results is shown in Figure 2.17.

As expected, at low flowrates the increased friction caused by wear and limescale influences metering errors, with values close to -35% for meters with accumulated volumes of $7500\,\text{m}^3$. In low flows, the obstruction of the regulation circuit does not compensate this increase in friction. The starting flowrate is also increased with age.

The rest of the tested flowrates present a rate of decay of the error completely opposite to that obtained at 30 l/h. In these cases, the errors increased lineally but with a positive slope (and with larger errors for larger flowrates). And so, with an accumulated volume of 7500 m^3 the average error of the meters tested at 1500 l/h was almost of +10% and a considerable number of these presented errors above +20%.

This phenomenon may present serious inconvenience for the service provider, as users could end up being charged a volume significantly larger than the one really consumed.

2.2.7. Metrology of multiple jet meters: Reverse flow

Similarly to what happens with single jet meters, the error in multijet meters depends on the impact velocity of water on the impeller, and consequently for a given flow, on the cross section of the entry to the distribution chamber. The cross section, and consequently the impact velocity on the impeller, is different when the water enters the meter from above (reverse flow, Figure 2.16) for a given flowrate. This causes the error curve to be different depending on the direction of the flow.

Figure 2.18 shows part of the error curve for a meter with a nominal flow of 3.5 m^3/h for both forward and reverse directions of the flow. In the regular direction, the error is approximately in the +2% range, but for reverse flows, the negative error is taken to −25%. These figures are only indicative, as the change in magnitude of the error will be a function of the difference in cross sections for the entry and exit to the distribution chambers, which vary depending on the specific model tested.

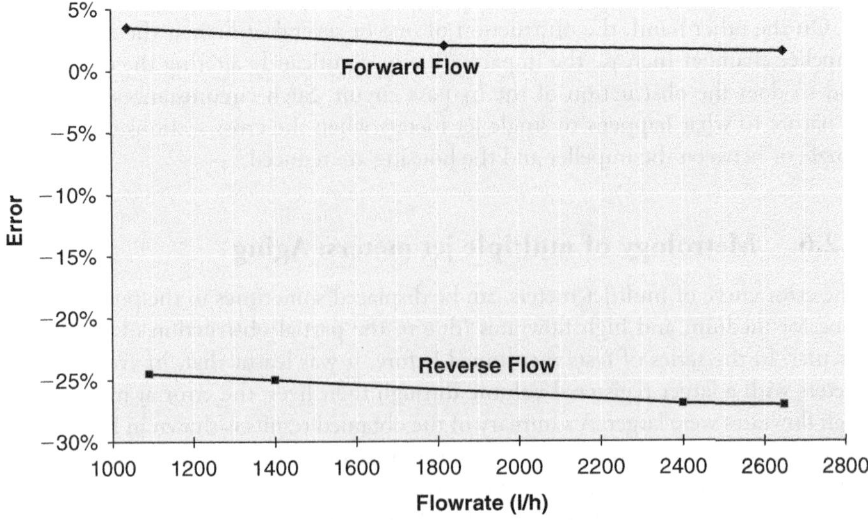

FIGURE 2.18. ERROR OF A MULTIJET METER AS A FUNCTION OF FLOW DIRECTION

2.2.8. Advantages and disadvantages of multiple jet meters

Advantages	Disadvantages
• Very reliable metering technology, used for decades around the world. • Forces on the impeller are balanced, providing a long life for the meter. • Not sensitive to the velocity profile at the entrance. The pipe preceding the meter does not need to be straight. • Good resistance to suspended solids. Adequate for hard water or water with suspended particles. • Great variety of models, construction, sizes and prices. An option for almost every use can be found. • Meters ranging from 20 to 40 mm are very competitive in price compared to other technologies.	• Larger than single jet meters, especially for small diameters. • Affected by the installation position. • Error curve is often displaced to positive errors for medium and large flows. Potential conflict with users. • Above 50 mm, their flow registering capacity is much smaller than the Woltmann's. • There is no Class D (ISO 4064:1993) model known to date. Class C meters above 30 mm are rarely found. • Starting flowrates are not low enough to detect most leaks at the users' side. Depending on the quality, this parameter can deteriorate quickly.

2.3. OSCILLATING PISTON METERS

2.3.1. Operating principles

The oscillating piston meters are, for their operating principle, positive displacement meters. They are mainly used for domestic uses and are quite common in some parts of the world.

These meters register the volume by counting the number of times a chamber of a known volume is filled and emptied. The moving part of the meter is a rotating piston with an eccentric motion around the metering chamber axis. Figure 2.19 shows how the two compartments of the metering chamber are filled and emptied. The entry orifice is situated at the side of the division plate. Due to a higher pressure in the upstream side, the piston tends to rotate (eccentrically guided). The division plate enters the piston at different degrees through an opening designed to that purpose.

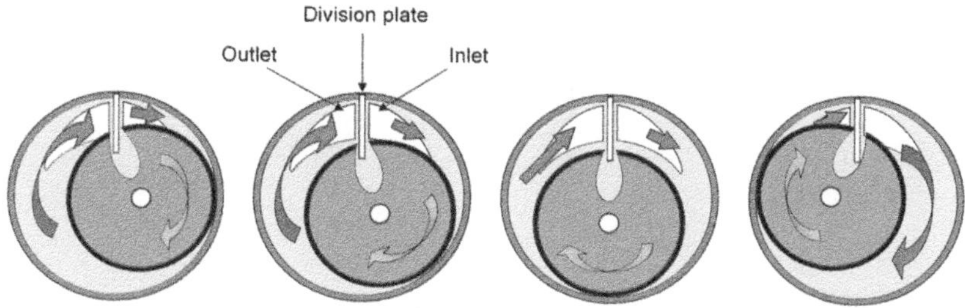

FIGURE 2.19. OPERATING PRINCIPLE OF AN OSCILLATING PISTON METER

As the compartment on the right is filled, the compartment situated to the left is emptied. In each rotation of the piston, the same volume of water enters and leaves the metering chamber (without considering possible leakage between the two compartments).

This technology presents the best accuracy to meter water volumes. As a matter of fact, the only Class D meters in the market are oscillating piston meters.

2.3.2. Constructive characteristics

An oscillating piston meter (Figure 2.20) is basically formed by:

- *The metering set*: integrated by the metering chamber, the piston and the division plate. The set is usually made out of plastic, although sometimes metals are also used.
- *Housing*: Usually made of brass or bronze, although sometimes plastic is used.
- *Strainer*: A fundamental part of this technology. It avoids any particles suspended in the water entering the metering chamber obstructing the moving parts of the meter.

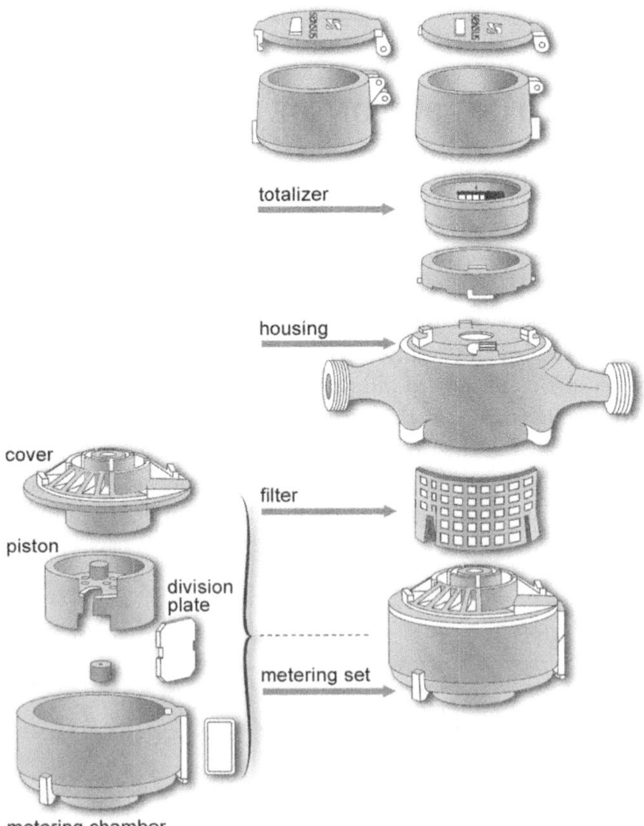

FIGURE 2.20. PARTS OF AN OSCILLATING PISTON METER (COURTESY OF SENSUS)

The accuracy of the meter mainly depends on the adjustments between the chamber and the piston, since metering errors originate in the leaks in that area. Consequently the tolerances in the dimensions of these two elements are critical. Excessive wear may lead to negative errors. On the other hand, a reduction in the inner volume of the chamber, for instance due to limescale, will not allow the rotation of the piston and consequently the operation of the meter.

In order to avoid the interruption of the supply, should the piston get jammed, some manufacturers drill holes in the piston in such a way that the entry and exit of the metering chamber are communicated. This method allows a minimum supply, although the meter will need to be either repaired or replaced on short notice. This solution, however, will make negative errors increase, especially at low flows when friction's drag torque plays an important role.

Some manufacturers have introduced modifications in the design of the pistons by including grooves in their surface. A grooved piston creates eddies that improve the characteristics and durability of the meter by preventing the adherence of particles to the surfaces of piston and chamber (Figure 2.21).

Another common source of complaints by users, mainly in older oscillating piston models, was to do with the noise that the piston produces in contact with the walls of the metering chambers. However, the use of better materials and the improvement in the manufacturing processes have significantly reduced this problem.

Oscillating piston meters can be classified in two types of constructions, depending on the rotation axis of the piston. In some of them, the rotation of the piston takes place parallel to the direction of the pipe. This is a typical solution adopted in the United Kingdom and Latin America (Figure 2.22, left). This type of construction provides smaller meters, an important factor at the time of installation. However, there is an important disadvantage in the limited amount of options provided by the totalizer. Additionally, most of these meters have a mechanical transmission of the motion between the piston and the totalizer, which may be the origin of severe metering problems. The typical metrological class for this construction is Class B.

The other type of oscillating piston meter is characterized by the rotation of the piston around an axis perpendicular to the direction of the pipe (Figure 2.22, right). This option is the most common in continental Europe and North America, and the meters can be found in both C and D Classes. The main advantage to this arrangement is that the totalizer used is the same one as in other types of meters, making the usage of pulse emitters, flow indicators, etc., possible.

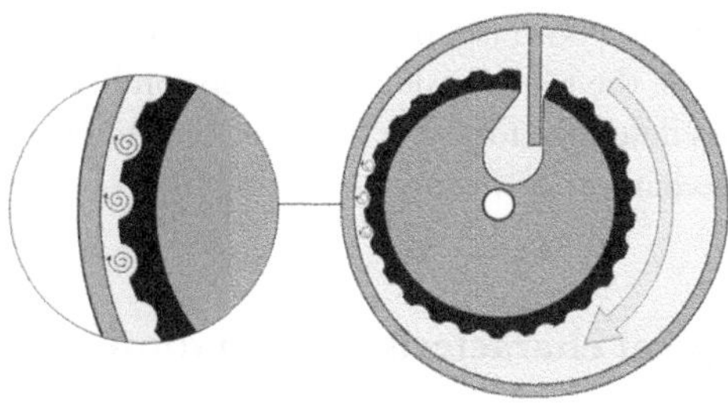

FIGURE 2.21.   GROOVED OSCILLATING PISTON

FIGURE 2.22. DIFFERENT CONFIGURATIONS OF AN OSCILLATING PISTON METER [COURTESY OF ELSTER (LEFT) AND ACTARIS (RIGHT)]

As has been mentioned before, one of the key elements for the adequate operation of a volumetric meter are the strainers. Without them, the piston would get blocked soon after installation due to the particles suspended in the water.

There are different possibilities when it comes to installing a strainer. A flat strainer can be arranged at the entry of the metering set and a second strainer such as a thimble strainer can be placed at the entry of the meter. In any case, the objective is to eliminate any possibility of foreign objects entering the metering chamber, while minimizing the pressure loss originated by the strainers. As a consequence, if the mesh of the strainer needs to be small, the cross section of the strainer will necessarily have to be larger. Otherwise, the pressure loss would be too elevated.

However, and despite the fact that strainers are essential elements, they require maintenance and an appropriate design. For instance, a study carried out by the authors in a Latin American city showed that oscillating piston meters (rotation parallel to the pipe) of length 115 mm were more prone to stop than those of length 190 mm. The explanation of this phenomenon was simple, and lay in the fact that the shorter meters had little space between the body of the meter and the strainer. As a consequence, the particles were trapped in this space and the strainer was quickly blocked. Curiously, in this case, users had domestic tanks in their properties and in many cases did not notice the small flows going through the meter and refilling the tanks. However, frequently, these flows were below the starting flowrate and were not metered or billed by the company.

Finally, and regarding the transmission of the motion between the piston and the totalizer, the principles described for other metering technologies apply, and the coupling can be either mechanical or magnetic, although the later option is, by far, the most frequent.

2.3.3. Metrological characteristics and dimensions

Oscillating piston meters are generally made for domestic use and in small and medium calibres. Nevertheless, in some countries it is possible to find models of larger

TABLE 2.7. METROLOGICAL CHARACTERISTICS OF CLASS C OSCILLATING PISTON METERS

Diameter (mm)	15	15	20	20	25	30	40	65
Nominal flowrate (m^3/h)	1	1.5	1.5	2.5	3.5	5	10	20
Maximum flowrate (m^3/h)	2	3	3	5	7	10	20	40
Transitional flowrate ($\pm 2\%$) (l/h)	15	22.5	22.5	37.5	52.5	75	150	300
Minimum flowrate ($\pm 5\%$) (l/h)	10	15	15	25	35	50	100	120
Starting flowrate (l/h)	1	1	2	2	6	11	18	30
Pressure loss at maximum flowrate (bar)	1	1	1	1	1	1	1	1

dimensions. Table 2.7 shows the main metrological characteristics of these meters as presented by the manufacturers.

It is important to note the high sensitivity of these meters for low flows, with starting flowrates much lower than the other technologies (generally speaking, the starting flowrate of a volumetric meter of $1.5\,m^3/h$ of nominal flowrate will range from 1 to 5 l/h). As a consequence, volumetric meters are able to detect leakage in the users' facilities, even dripping taps.

Volumetric meters are therefore adequate wherever consumption flows are low, the price of water is high and/or there is a high volume of leakage in the user's property. These meters are not recommended for users with seasonal water consumption for which the meters are stopped during a considerable period of time.

2.3.4. Metrology of oscillating piston meters: Orientation

Unlike velocity-based meters, volumetric meters are hardly affected by the installation position and the error curve remains almost constant for all possible orientations. The main divergences appear at low flows and are due to an increase in the filtrations between the chamber and the piston caused by higher friction in the mobile parts. In any case, these differences are not significant for most applications.

Most manufacturers in their installation instructions allow all orientations for the meter, while ensuring that the metrological class will not be lost in the process.

2.3.5. Metrology of oscillating piston meters: Velocity profile

The operating principle of volumetric meters guarantees that they will not be affected by changes in the velocity profile. Consequently, the presence of any kind of element upstream from the meter will not affect its error curve.

Additionally, a partial blockage of the strainer will not affect the quality of measurement, as long as water is able to flow through it. This particular aspect was studied by the authors in a Latin American city, where a sample of volumetric meters with dirty strainers was tested, both with and without strainer. The results, as expected, showed that there were no significant differences between the error curves of the meters in both tests.

2.3.6. Metrology of oscillating piston meters: Aging

Unlike single and multiple jet meters, there are few things that can affect the performance of a volumetric meter. Overregistration will only take place if there is a reduction

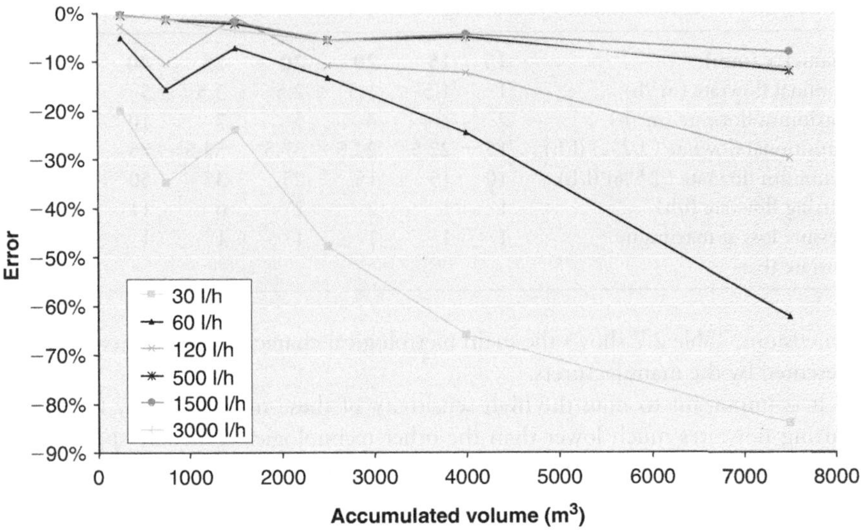

FIGURE 2.23.　EVOLUTION OF THE METERING ERROR OBTAINED FROM A SAMPLE OF VOLUMETRIC METERS (½″)

of the actual volume of each compartment. This is virtually impossible, for this reduction would make the meter stop. Consequently, it can be stated that the error curve of such meters always has a tendency to become more negative (underregister) with respect to the original one.

As already mentioned before, in these cases the inaccuracies come mainly from the increase in the torque resisting the motion and the tolerances between the chamber and the piston, which leads to a higher-volume leaking through both compartments. As a consequence, as the meter gets older, for every oscillation of the piston, the amount of water flowing through the meter is larger. This is the volume of water that is "lost" and not registered, and as a consequence there is a displacement of the error curve towards negative values.

Figure 2.23 shows the evolution of the error curve for a sample of 250 ½″-volumetric meters with different accumulated volumes.

Several conclusions may be drawn from the previous figure:

- As already explained in theory, the metering error in volumetric meters has only a tendency to negative values.
- The metering error at low flowrates is higher than the error found at higher flowrates.
- The higher the flowrate, the longer the meter stays accurate with time.
- For 1.5 m³/h meters the downgrade of the error from flowrates of 500 l/h onwards is basically the same.

2.3.7.　Metrology of oscillating piston meters: Reverse flow

Volumetric meters assess the consumption by counting the number of times that the compartments of the metering chamber are filled and emptied. Consequently, the error curve of the meter is almost the same, regardless of the direction of the flow. In both cases the compartments are filled and emptied completely, and the only difference that could take place would be an increase in friction leading to a higher-volume leaking between the chamber and the piston (water flowing from one compartment to the other).

2.3.8. Metrology of oscillating piston meters: Entrapped air

The volumetric meters are also known as positive displacement meters. Due to this operating principle, if a fluid other than water occupies part of the volume in the metering chamber (for instance, air) the metering error can be quite severe and rapid wear of mobile parts may take place.

To avoid these sort of errors and, as it happens with other types of meters, it is recommended that an air valve is installed in an elevated point of the pipe upstream from the meter.

2.3.9. Metrology of oscillating piston meters: Foreign objects

The particles suspended in the water can severely modify the error curve for these meters. This is possibly one of the main reasons behind the reticence of many undertakings to use this metering technology. It is quite difficult to guarantee, due to the frequent repairs in the network, that the quality of the water reaching the users will be of high enough quality for volumetric meters to operate correctly.

The use of strainers is strictly necessary. However, strainers may not be enough and the problems and complaints from the users may be too many compared to the benefits offered by this technology. Furthermore, strainer maintenance will increase the maintenance costs with respect to other metering technologies.

2.3.10. Advantages and disadvantages of oscillating piston meters

Advantages	Disadvantages
<ul><li>Very reliable technology used for decades.</li><li>Not affected by the velocity profile. Less space needed for their installation.</li><li>Great variety of models, metrologies and prices.</li><li>Excellent sensitivity at low flowrates.</li><li>Available in Class D.</li><li>Almost unaffected by the installation position.</li><li>Error is always on the negative side of the curve (underregistration) avoiding problems with users.</li></ul>	<ul><li>Affected by suspended solids. This is the main argument posed by users for not using them.</li><li>Noisy at high flowrates.</li><li>Larger and heavier than equivalent meters of other technologies.</li><li>More expensive than equivalent meters of other technologies.</li></ul>

2.4. NUTATING DISC METERS

2.4.1. Operating principles

The nutating disc meters are by principle volumetric meters. Just like the oscillating piston meters they are used for domestic metering and are common in many countries, especially in North America.

The consumption is determined by counting the number of cycles in which a chamber of known volume is filled and emptied. The moving element is a disc that oscillates around the symmetry axis of the chamber, dividing it in separate compartments. The water enters the chamber from one side, filling up one of the compartments. The disc rotates eccentrically to simultaneously perform a number of tasks: closing the entrance to the chamber, revealing the exit to the chamber and reducing the volume of the compartment. The same compartment never witnesses both orifices (entry and exit) open at the same time. Consequently, in each rotation of the disc, the same volume of water passes through the meter.

2.4.2. Constructive characteristics

In these meters, the metering device is formed by the metering chamber, the disc and the division plate. It is usually manufactured in plastics although sometimes metals are also used. The metering chamber has a peculiar shape in order to adapt to the oscillating motion of the disc (Figure 2.25). The upper and lower parts have conical shapes, while the sidewalls are curved (spherical).

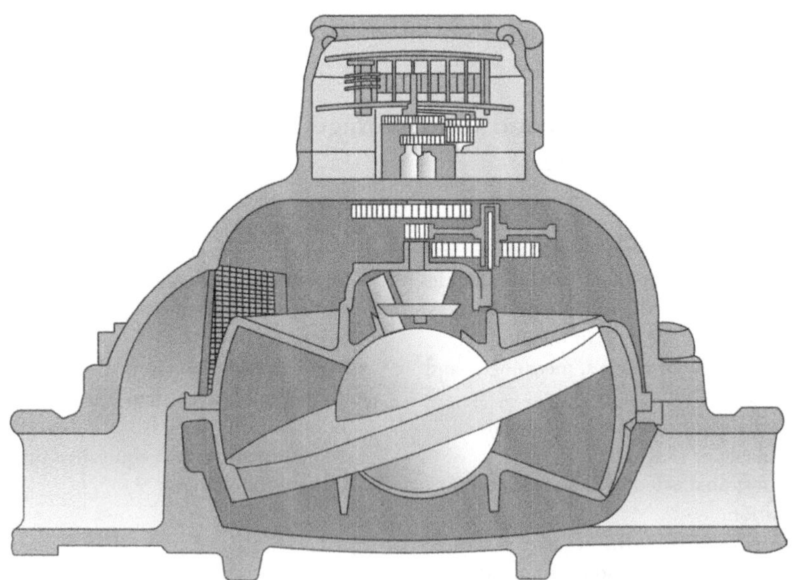

FIGURE 2.24. CROSS SECTION OF A NUTATING DISC METER

The housing of the meter contains the metering device and the strainers. It is usually built in brass or bronze, although sometimes plastics are used.

Just like in oscillating piston meters, the quality of water plays a key role in the durability and metrological performance of the meter. As a consequence, strainers are absolutely necessary in order to avoid suspended particles that could block the disc's motion.

The metering errors mainly depend on the adjustment between the chamber and the disc. The leakage in these elements causes the difference between the registered and the actual volume flowing through the meter. Consequently, the dimensions and tolerances in these elements are critical. Excessive wear in the parts will lead to negative errors in metering.

The transmission of the disc's motion to a vertical shaft is always done mechanically. However, as with all other meter technologies, the transmission to the totalizer's gear can also be done magnetically.

2.4.3. Metrological characteristics and dimensions

As mentioned before, since nutating disc volumetric meters are mainly manufactured in the United States for domestic use (generally, in small and medium calibres), and the international standard ISO 4064 is not commonly applied in this country, the metrology of these meters is dictated by the American standards C700-02 and C710-02. Table 2.8 shows the main metrological characteristics as provided by one manufacturer.

As opposed to ISO 4064, the AWWA standard pays special attention to the accuracy of the meter at high flowrates. This is the reason why, although the measuring range is very wide, the sensitivity at low flowrates is not as good as it would be with an ISO certified meter.

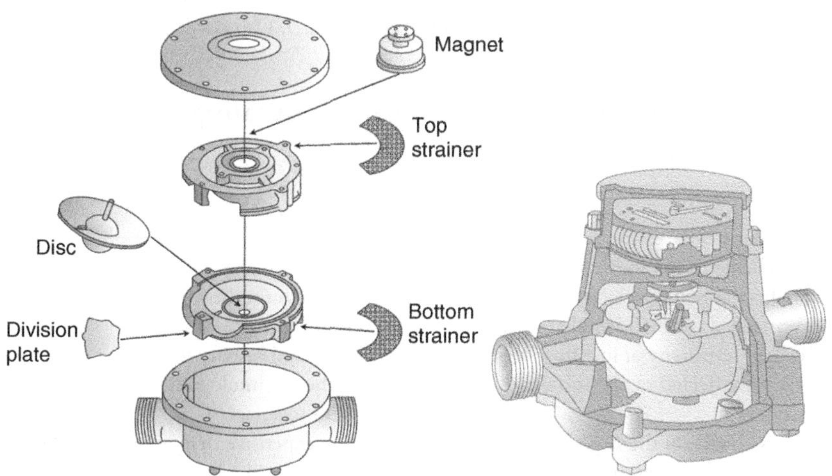

FIGURE 2.25. PARTS SCHEMATIC OF A NUTATING DISC METER

TABLE 2.8. METROLOGICAL CHARACTERISTICS OF NUTATING DISC METERS (AWWA C700)

Diameter (mm)	5/8″	¾″	1″	1½″	2″
Safe maximum operating capacity ($\pm$1.5%) AWWA (m³/h)	4.5	6.8	11.4	22.7	36.3
Actual safe maximum operating capacity ($\pm$1.5%) (m³/h)	4.5	6.8	11.4	22.7	36.3
Minimum flowrate ($\pm$1.5%) at normal operating range AWWA (m³/h)	0.23	0.45	0.68	1.1	1.8
Actual minimum flowrate at normal operating range ($\pm$1.5%) (m³/h)	0.11	0.17	0.23	0.46	0.57
Actual minimum flowrate ($-$5%) (m³/h)	0.03	0.06	0.09	0.17	0.23
Pressure loss at maximum flowrate (bar)	0.55	0.69	0.55	0.69	0.69

As an example, a ½″ oscillating piston meter (Class C ISO 4064:1993) and a 5/8″ nutating disc meter (AWWA C700) can be compared. The maximum flowrate of the first one is 3 m³/h, while the AWWA approved meter can go all the way up to 4.5 m³/h (50% more). However, the minimum flowrate ($\pm$5%) of the Class C meter is 15 l/h while the standard AWWA meter reaches this accuracy at 60 l/h (1/4 gpm).

In any case, when compared by flowrate capacity and Class B, the differences are not as noticeable. In this case, a Class C ISO meter with a nominal rate of 2.5 m³/h, maximum flowrate of 5 m³/h and minimum flowrate of 50 l/h, would be equivalent in performance, although slightly superior to a 5/8″ AWWA C700 meter. ISO4064:1993, Class C and B requirements are always stricter than AWWA's.

2.4.4. Metrology of nutating disc meters

The external factors affecting the metrology of nutating disc meters are the same as for other volumetric meters, and can be found in Section 2.3.4 for oscillating piston meters.

2.4.5. Advantages and disadvantages of nutating disc meters

Advantages	Disadvantages
• Very reliable technology used for decades. • Not affected by the velocity profile. Less space needed for their installation. • Excellent sensitivity at low flowrates. • Almost unaffected by the installation position. • Error is always on the negative side of the curve (underregistration) avoiding problems with users.	• Affected by suspended solids. This is the main argument posed by users for not using them. • Larger and heavier than equivalent meters of other technologies, even oscillating piston meters. • More expensive than equivalent meters of other technologies.

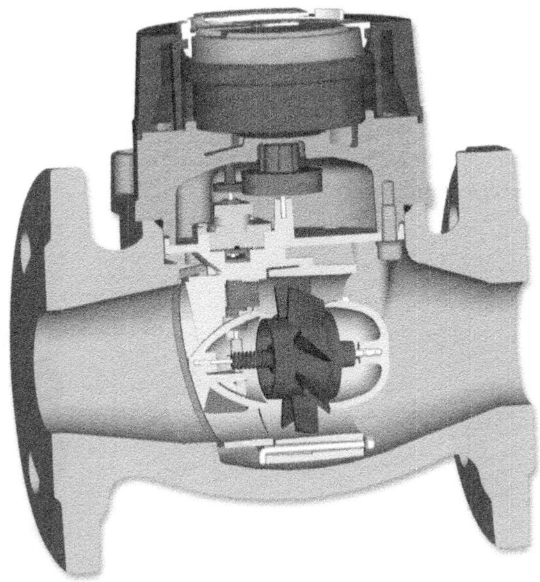

FIGURE 2.26.   CROSS SECTION OF A WOLTMANN METER (COURTESY OF ACTARIS)

2.5.   WOLTMANN METERS

2.5.1.   Operating principles

Woltmann meters get their name from the German engineer Reinhard Woltmann (1757–1830) who introduced the use of helices in open channel metering.

The primary element of a Woltmann meter is a wheel facing the flow in its axial direction. The rotational speed of this wheel is a function of the flowrate, which determines the impact velocity of water on the blades, and the design of the blades and their angle (Palau et al., 2004; Baker, 2000). This is a key issue, since the relative impact angle of the water on the blade of a given Woltmann meter is a function of the velocity profile, and consequently an improper installation of the meter will affect its performance significantly (especially when swirly flow is generated upstream the meter).

As a general rule of thumb, a meter design with a steeper blade angle will allow for lower starting flowrates, since the transmission of angular momentum to the wheel for a given velocity is larger. However, this will also create higher rotation speeds and could lead to a higher wear of bearings and gears.

2.5.2.   Constructive characteristics

Woltmann meters are extensively used around the world for diameters between 50 and 500 mm. Exceptionally and using special housings and constructions, they can be found in diameters up to 800 mm.

Taking into account the rotation axis of the wheel, there are three possible configurations for a Woltmann meter: vertical vane, horizontal vane and angle (Figure 2.27). However, the horizontal vane Woltmann meter is by far the most common one.

Horizontal vane Woltmann meters

For this configuration, the rotation axis of the meter wheel is parallel to the direction of the flow. The housing of the meter does not significantly affect the velocity profile at the entry of the meter and consequently the characteristics of the incoming flow determine its performance. Figure 2.28 shows a cross section of a horizontal vane Woltmann meter with its different components.

FIGURE 2.27.   CONFIGURATIONS OF A WOLTMANN METER (HORIZONTAL AND VERTICAL COURTESY OF ACTARIS – ANGLE COURTESY OF SENSUS)

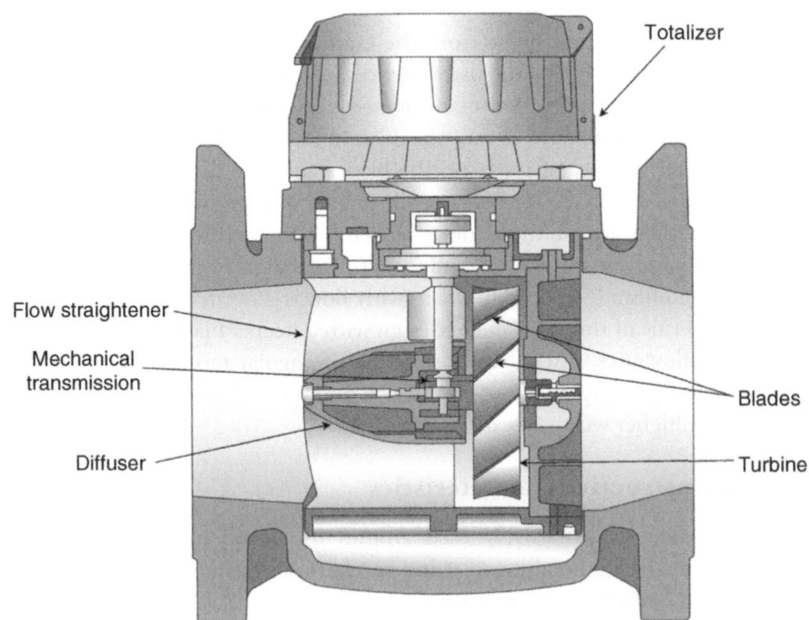

FIGURE 2.28.   CROSS SECTION OF A HORIZONTAL VANE WOLTMANN METER (COURTESY OF ELSTER)

In the meter shown in Figure 2.28, the wheel axis is parallel to the flow. However, the rotational axis of the totalizer's gears is perpendicular to this direction. The transmission of the rotational motion from one axis to a perpendicular one creates a considerable drag torque which reduces the meter sensitivity to low flows. For this reason, amongst others, there are no Class C horizontal vane Woltmann meters.

Horizontal vane Woltmann meters can be found in the whole diameter range (50–800 mm) although the design for the larger-diameters changes and the metering element does not occupy the whole transversal section (Figure 2.29).

The regulation of the meter's error curve is achieved by modifying in a region of the cross section upstream the turbine the relative incidence angle of flow and the blades, using a regulation plate like the one shown in Figure 2.30. The orientation of the plate modifies, in the area near it, the angle at which the flow impacts on the wheel blades, changing the driving torque and, as a consequence, the rotational speed for a given flowrate. A reduction in the rotational speed of the wheel will displace the error curve towards negative values.

In order to use Woltmann meters in the field, it is important to note that a small element like the regulation plate is capable of displacing the error curve up to a $\pm 5\%$, depending on manufacturer and model. It is easy to imagine that larger perturbations, such as the ones created by regulation valves, can have drastic consequences on the performance of the meter.

FIGURE 2.29. LARGE HORIZONTAL VANE WOLTMANN METER (COURTESY OF SENSUS)

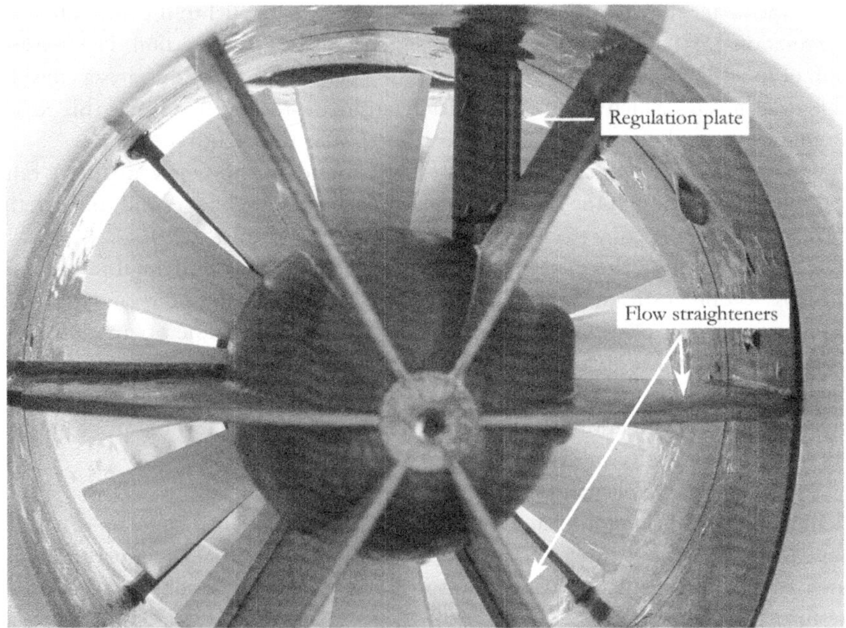

FIGURE 2.30. HORIZONTAL WOLTMANN METER WITH REGULATION PLATE

As a consequence, it is of the greatest importance, that all fluid layers reach the meter without any rotational motion. Only then, will the performance of the meter remain within the specifications defined for its class. Any kind of swirl introduced in the fluid motion will greatly affect the accuracy of the measure. Other distortions of the velocity profile, such as the lack of axial symmetry, are also to be avoided.

In order to prevent these flow distortions, Woltmann meters usually incorporate flow stabilizer plates at the entry of the meter. These plates, which can be seen in Figure 2.30, are usually made of plastic and prevent the helicoidal motion of water when reaching the wheel. The wider these elements are in the direction of the flow, the better they will stabilize the flow, and consequently will derive in a better measure. However, the capacity of the plates to stabilize the flow is limited, and quite often external stabilizing devices are needed in order to regularize the velocity profile.

Just like single and multiple jet meters, the transmission of motion between the wheel and the totalizer can be done mechanically or magnetically (which is the current choice for most manufacturers given the advantages already commented in this chapter).

Woltmann meters are mostly used to register consumption from large users or even to measure the injected volumes in small and medium networks. For these same reasons, quite often they incorporate different pulse emitters in order to determine peak and minimum flowrates, the hourly registered volumes, etc.

Figure 2.31 shows the schematics for the totalizer of a horizontal vane Woltmann meter which can be fitted with several types of pulse emitters (two Reed type emitters

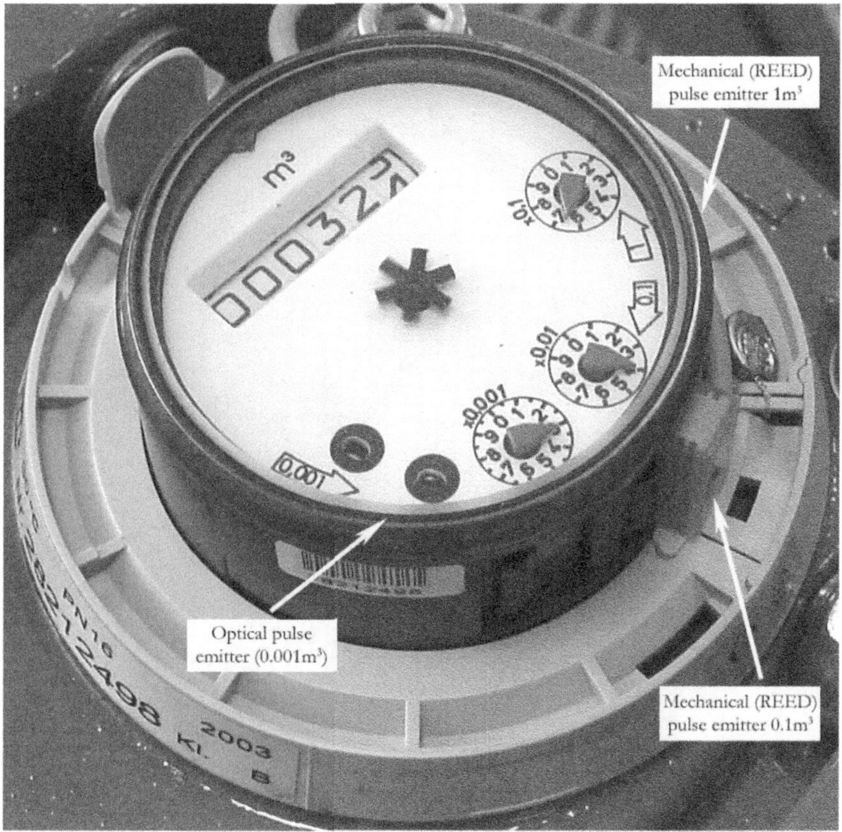

FIGURE 2.31. PULSE EMITTERS IN A HORIZONTAL WOLTMANN METER

of 0.1 and 1 m^3, respectively and an optical one). Depending on the gear used to obtain the signal, the volume corresponding to each pulse will be different.

High-frequency signals, with a small associated volume, are characteristic of photoelectric or inductive pulse emitters. On the other hand, low-frequency signals, with a pulse for every hundred or thousand litres, are usually found in REED type emitters. This is due to the fact that REED type meters are limited in frequency to 1 Hz, and lower volumes per pulse could saturate the emitter, leading to loss of information.

The integrated metering module (wheel, flow stabilizers and totalizer) of many Woltmann meters can be replaced (Figure 2.32). In theory, the replacement set is homologated, and the quality of the measure is not affected by the replacement. The replacement operation is simple, and it can be performed in a few minutes with no need of especial tools. The cost of the replacement module is lower than the cost of acquiring and installing a full meter. This maintenance possibility represents an important advantage of Woltmann meters over other technologies such as single jet or electromagnetic meters.

FIGURE 2.32.   REPLACEMENT OF THE METERING MODULE IN A WOLTMANN METER

Vertical vane Woltmann meters

The wheel of a vertical vane Woltmann rotates perpendicularly to the direction of the pipe. In order to achieve this, the direction of the flow has to be changed and made parallel to the rotation axis of the wheel (Figure 2.33).

This design introduces alterations in the velocity profile impacting the turbine. For this same reason, the distortions of the flow that may have happened upstream of the meter will have a much lesser effect on its accuracy. Consequently, vertical vane Woltmann meters are not greatly affected by the characteristics of the incoming flow and the length of straight pipe needed upstream of the meter is much shorter (Figure 2.37).

As shown in Figure 2.33, water enters a vertical vane Woltmann meter parallel to the direction of the pipe, to later descend to a distribution chamber placed in the lower part of the meter. In this chamber another change of direction is made to let the water flow parallel to the turbine's axis. Finally, the flow direction is again changed to allow water to leave the meter in the original direction.

The main advantage of this configuration is the smaller drag torque of the turbine, allowing for a greater sensitivity of the meter at low flowrates (compared to equivalent horizontal vane Woltmann meters). The reduction in the torque is a consequence of the

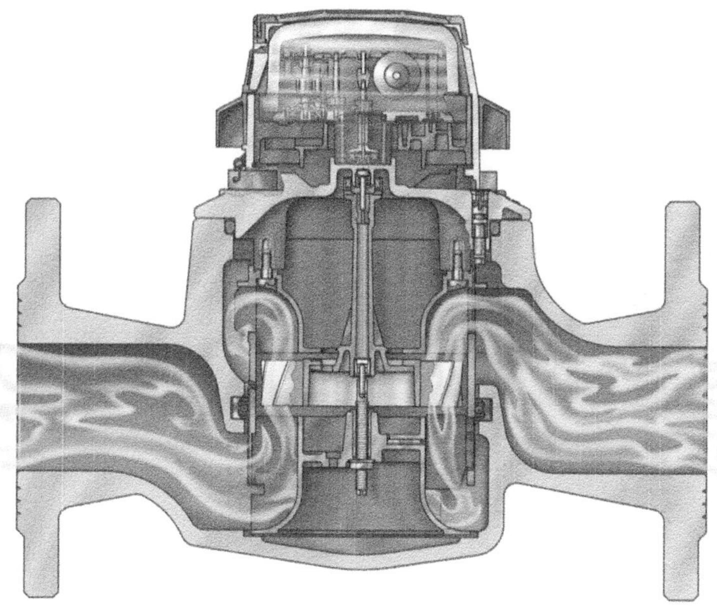

FIGURE 2.33. CROSS SECTION OF A VERTICAL VANE WOLTMANN METER (COURTESY OF SENSUS)

alignment of the axis of the turbine and the totalizer gears, avoiding the need for mechanisms that change the direction of the motion incrementing the resisting torque.

Vertical vane Woltmann meters are usually manufactured in Class B and with mechanical totalizer, although models with electronic totalizer can be found up to Class C.

The vertical position of the turbine presents also some disadvantages. The longer and winding path of the fluid inside the meter housing increases the friction losses. A horizontal vane Woltmann presents pressure losses at maximum flowrate ranging from 0.1 to 0.25 bar. In a vertical vane Woltmann, these values range from 0.4 to 0.6 bar. This may be an important fact when taking into account the total cost of the meter, which should include also operation costs, such as the dissipated energy. Additionally, the maximum flow in a horizontal vane Woltmann is much higher than in a vertical vane one for the same nominal diameter.

The adjustment of the error curve of the meter is done in the same way as in a horizontal vane Woltmann, by means of a regulation plate that provides a slight rotation to the fluid before reaching the turbine.

Finally, in vertical vane Woltmann meters, it is also possible to replace the metering module with all the advantages already mentioned before in terms of the maintenance costs and quality of the measure.

Angle Woltmann meters

Woltmann meters with a 90° vane are mainly designed for underground water abstraction. Their main characteristic is that the flow bends in a 90° angle from entry to exit,

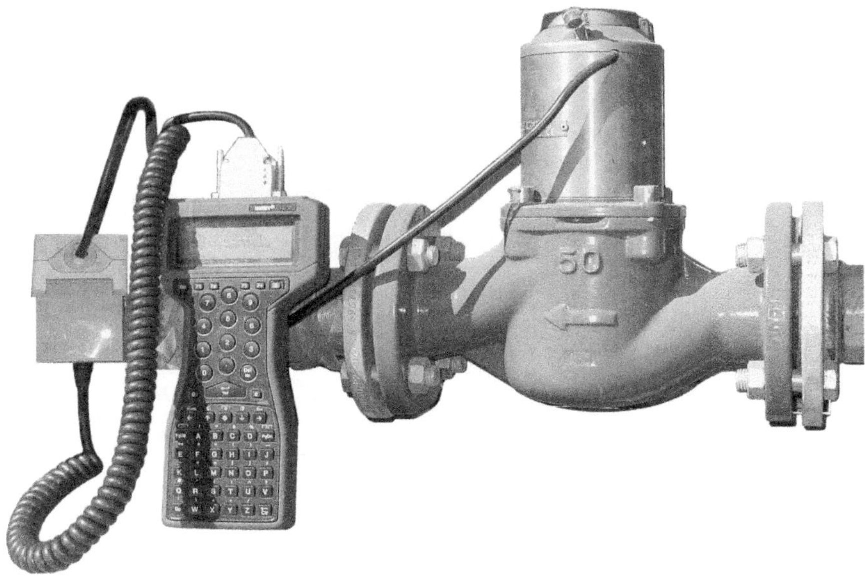

FIGURE 2.34. CLASS C VERTICAL VANE WOLTMANN METER WITH ELECTRONIC TOTALIZER (COURTESY OF CONTAZARA)

and consequently installing the meter avoids the additional costs of setting up an elbow joint in the pipe.

The axis of the turbine of a 90° Woltmann is placed in the direction of the flow at the entry, and is also parallel to the totalizer rotation axis. The water enters the meter from below, where it faces flow stabilizers, finally reaching the turbine. Afterwards, the flow changes in direction and finally exits the meter (Figure 2.35).

The housing of the meter does not affect the velocity profile before it reaches the turbine. As a consequence, the meter is as sensitive to the incoming velocity profile as a horizontal vane Woltmann. The adjustment of the error curve is done through a regulation plate, and angle Woltmann meters can only be found as Class B.

Pressure losses in angle Woltmann meters are slightly lower than in vertical vane models, while maximum flow capacity is almost the same (since the design is almost identical).

2.5.3. Metrological characteristics and dimensions

The metrological and dimensional values for all types of Woltmann meters provided here are dependant on the manufacturer, and consequently may change for specific models.

Horizontal vane Woltmann meters

The real metrological characteristics provided by the manufacturers are above the requirements of the corresponding class, even by a wide margin occasionally. As a

consequence, meters are often described both by their legal metrological characteristics and the real ones. The example of the 50 mm meter described in Table 2.9 is quite clear:

- The transitional flowrate is $5 \, m^3/h$. However, the meter actually reaches the $\pm 2\%$ range with only $0.7 \, m^3/h$. This is still however below the Class C requirements (in this case $0.375 \, m^3/h$).
- The minimum legal flow is $0.75 \, m^3/h$ while the real one is only $0.5 \, m^3/h$. For a Class C meter it would be of only $0.15 \, m^3/h$.
- The maximum required flowrate is $50 \, m^3/h$. However, the manufacturer "guarantees" this model for a few minutes at $90 \, m^3/h$.

The information provided by the manufacturer includes, as for many large calibre meters, additional information for choosing a meter of the right size. These include the recommended maximum monthly volume that the meter may register without deterioration, and is obtained by the number of hours in a month (720) and the permanent or nominal flow (Table 2.10).

Another important variable to be taken into account is the magnitude of pressure losses induced by the meter. It is a measure of the energy dissipated by the meter and its operating costs. The pressure losses in a meter are not constant and as all friction

TABLE 2.9. CHARACTERISTICS OF CLASS B HORIZONTAL VANE WOLTMANN METERS (I)

Diameter (mm)	50	65	80	100	150
Nominal flowrate (ISO 4064:1993) (m^3/h)	25	40	60	100	250
Peak flowrate (few minutes) (m^3/h)	90	200	250	300	700
Maximum flowrate (ISO 4064:1993) (m^3/h)	50	80	120	200	500
Maximum monthly volume (m^3)	18,000	28,800	43,200	72,000	180,000
Flowrate (10 h/day) (m^3/h)	37.5	60	90	150	375
Transitional flowrate (ISO 4064:1993) (m^3/h)	5	8	12	20	50
Real transitional flowrate ($\pm$2%) (m^3/h)	0.7	0.7	1.2	1.5	1.5
Minimum flowrate (ISO 4064:1993) (m^3/h)	0.75	1.2	1.8	3	7.5
Real minimum flowrate ($\pm$5%) (m^3/h)	0.5	0.5	0.75	1.2	1.8
Starting flowrate (l/h)	200	250	300	400	1400
Pressure loss at maximum flowrate (bar)	0.3	0.6	0.6	0.3	0.3

TABLE 2.10. CHARACTERISTICS OF CLASS B HORIZONTAL VANE WOLTMANN METERS (II)

Diameter (mm)	200	250	300	400	500
Nominal flowrate (ISO 4064:1993) (m^3/h)	400	600	1000	1500	2500
Peak flowrate (few minutes) (m^3/h)	1000	1500	2500	4500	7000
Maximum flowrate (ISO 4064:1993) (m^3/h)	800	1200	2000	3000	5000
Maximum monthly volume (m^3)	288,000	432,000	720,000	1,080,000	1,800,000
Flowrate (10 h/day) (m^3/h)	600	900	1500	2250	3750
Transitional flowrate (ISO 4064:1993) (m^3/h)	80	120	200	300	500
Real transitional flowrate ($\pm$2%) (m^3/h)	15	30	30	50	50
Minimum flowrate (ISO 4064:1993) (m^3/h)	12	18	30	45	75
Real minimum flowrate ($\pm$5%) (m^3/h)	4	6	15	25	30
Starting flowrate (l/h)	1600	3000	10,000	15,000	20,000
Pressure loss at maximum flowrate (bar)	0.3	0.3	0.3	0.3	0.3

losses are proportional to the square value of the flowrate. Manufacturers usually quote the value of pressure losses at a certain flowrate, either the maximum or the nominal.

The ISO 4064:1993 standard establishes the maximum pressure loss at 1 bar at the maximum flowrate. Single and multiple jet meters often reach values close to the limit. Horizontal vane Woltmann meters are normally between 0.3 and 0.4 bar, reaching values that even correspond to a more demanding class.

Vertical vane Woltmann meters

Vertical vane Woltmann meters are usually found in calibres ranging from 50 to 150 mm. Tables 2.11 and 2.12 show the characteristics provided by a manufacturer for meters of Classes B and C.

As mentioned before, the maximum flowrate for a vertical vane Woltmann meter is significantly lower than the one from an equivalent horizontal vane Woltmann. For instance the 100 mm vertical Woltmann, Class B, presents a peak flow of 180 m^3/h. The equivalent horizontal Woltmann presented before can reach the 300 m^3/h.

TABLE 2.11. CHARACTERISTICS OF CLASS B VERTICAL VANE WOLTMANN METERS (I)

Diameter (mm)	50	65	80	100	150
Nominal flowrate (ISO 4064:1993) (m^3/h)	15	25	40	60	150
Peak flowrate (few minutes) (m^3/h)	35	70	110	180	350
Maximum flowrate (ISO 4064:1993) (m^3/h)	30	50	80	120	300
Permanent flowrate (m^3/h)	20	40	55	90	250
Transitional flowrate (ISO 4064:1993) (m^3/h)	3	5	8	12	30
Real transitional flowrate ($\pm$2%) (m^3/h)	1	2.5	2.5	3	5
Minimum flowrate (ISO 4064:1993) (m^3/h)	0.45	0.75	1.20	1.80	4.5
Real minimum flowrate ($\pm$5%) (m^3/h)	0.15	0.2	0.2	0.3	0.8
Starting flowrate (l/h)	50	70	100	110	500
Pressure loss at maximum flowrate (bar)	0.23	0.27	0.38	0.28	0.58

TABLE 2.12. CHARACTERISTICS OF CLASS C VERTICAL VANE WOLTMANN METERS (II)

Diameter (mm)	50	65	80	100
Nominal flowrate (ISO 4064:1993) (m^3/h)	15	25	40	60
Maximum flowrate (ISO 4064:1993) (m^3/h)	30	50	80	120
Transitional flowrate (ISO 4064:1993) (m^3/h)	0.225	0.375	0.6	0.9
Minimum flowrate (ISO 4064:1993) (m^3/h)	0.09	0.15	0.24	0.36
Pressure loss at maximum flowrate (bar)	0.5	1	1	1

However, it has already been mentioned that vertical Woltmann meters are more sensitive to low flows, due to a lower friction in the transmission of the motion. Consequently the starting flow is much lower for a vertical vane Woltmann (110 l/h) compared to the equivalent horizontal vane model (300 l/h).

This increased sensitivity can also be noticed in the real flows that fall within the ±5% error range. A horizontal vane Woltmann meter will enter this band at 1.2 m³/h, while the vertical equivalent will obtain absolute errors less than 5% at 0.3 m³/h (almost the same value as a Class C). These differences, however, almost disappear when the error range considered is ±2%, as shown in the tables.

As a conclusion, vertical vane Woltmann meters do have a small advantage when considering low flows. However, the larger capacity of the horizontal vane models allows the selection of smaller calibres, for a given nominal flow, partly solving this problem. For instance, for a 60 m³/h, a probable choice of diameters would be 100 mm for a vertical vane model, and only 80 mm for a horizontal vane one. With this selection there are still differences in sensitivity, which in any case, have been reduced with the smaller diameter of the horizontal vane Woltmann.

Angle Woltmann meters

Angle Woltmann meters are not used very often for they are quite expensive. The range of diameters is very limited and spans from 80 to 200 mm. Table 2.13 shows the main characteristics provided by a manufacturer.

As a comparison between these values and the ones provided for vertical vane Woltmann meter, it can be stated that:

- Given equivalent diameters, the peak flowrate for an angle Woltmann is similar to the peak flowrate for a vertical vane one.
- The real flows at which the 90° meter reaches the ±5% and ±2% error ranges are higher. Consequently, it is less sensitive at low flows than the vertical vane Woltmann, although better than the horizontal vane ones.
- A similar trend can be found with the starting flows, which are higher than for vertical vane meters, but lower than those found in horizontal vane ones.

2.5.4. Metrology of Woltmann meters: Orientation

Horizontal vane Woltmann meters can be installed in virtually every orientation without the error curve being affected. The only limitation is the inclination of the

TABLE 2.13. CHARACTERISTICS OF CLASS B ANGLE WOLTMANN METERS

Diameter (mm)	80	100	150	200
Permanent flowrate (m³/h)	55	90	250	325
Peak flowrate (few minutes) (m³/h)	110	180	350	600
Real transitional flowrate (±2%) (m³/h)	3	6	15	30
Real minimum flowrate (±5%) (m³/h)	0.55	0.65	0.8	7.5
Starting flowrate (l/h)	200	300	500	3500
Pressure loss at maximum flowrate (bar)	0.2	0.25	0.40	0.50

totalizer which should never be placed upside down. Figure 2.36 shows different possibilities for the installation.

Vertical and angle Woltmann meters are very different and present serious restrictions to their installation position. The only allowed-for orientation is the meter being horizontal and the totalizer facing upwards. Any other position will make the turbine rest in the wrong position and will lead to early wear of the bearings and moving parts. Additionally, the low flow sensitivity of the meter will also be affected.

2.5.5. Metrology of Woltmann meters: Velocity profile

Horizontal vane Woltmann meters are very sensitive to the *quality* of the velocity profile reaching the turbine, especially in those cases where there may be radial components of the flow. The only way to ensure that the velocity profile is completely regular is to place enough length of straight pipe ahead of the meter or a flow stabilizer.

The length of straight pipe needed to regularize the velocity profile depends on the original perturbation and the meter itself. The distortion created by a bend in the pipe is not the same as the one created by two bends in different planes (Figure 2.37). The first case would require, from the recommendations of a manufacturer, a length of 12 diameters ahead of the meter, while the second one would require up to 20, an apparently excessive number.

It is however important to take into account that the recommendations change a lot from one manufacturer to the other. For instance, recommendations shown in

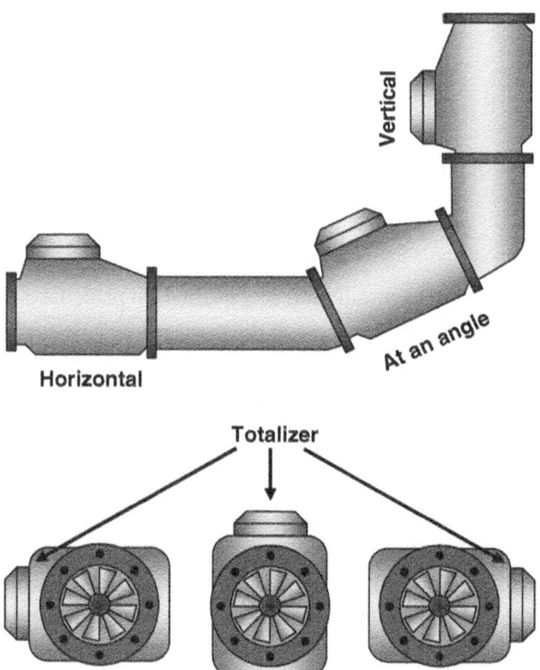

FIGURE 2.36. ALLOWED ORIENTATIONS FOR A HORIZONTAL VANE WOLTMANN METER

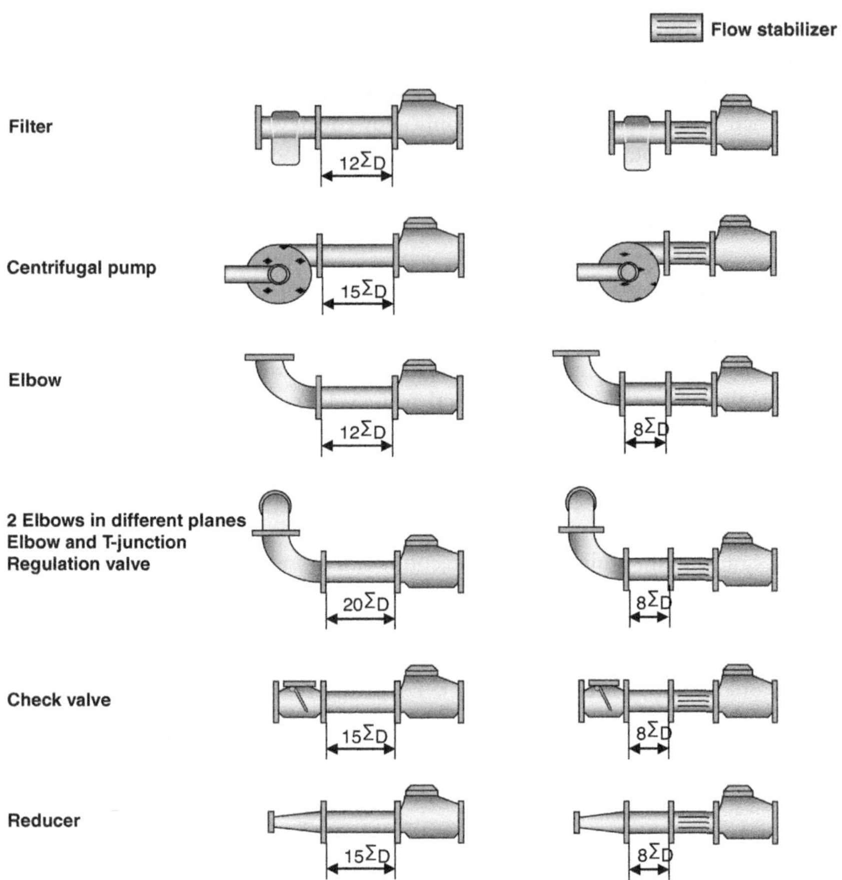

FIGURE 2.37. REQUIRED LENGTHS OF STRAIGHT PIPE AHEAD OF A HORIZONTAL VANE WOLTMANN METER

Figure 2.37 demand, after the installation of a strainer, to place 12 diameters of straight pipe. However, another manufacturer will only suggest 3–5 diameters, depending on the nominal diameter of the meter, for a similar perturbation. As a consequence, it is suggested always to follow the recommendations from the manufacturer of every specific meter.

Quite often, the diameter of the meter to use is smaller than the diameter of the pipe, creating the need for reductions. A common doubt in these cases is how these reductions will affect the measure. As a general rule, these reductions will not affect the velocity profile. They will only accelerate the flow, increasing all velocities without affecting their distribution. Consequently, and as long as the diameter change takes place gradually, the use of reductions requires no further action to provide accurate measures.

In case of having a distorting element upstream from the reduction, the recommendations of the manufacturer should be applied as shown in Figure 2.38.

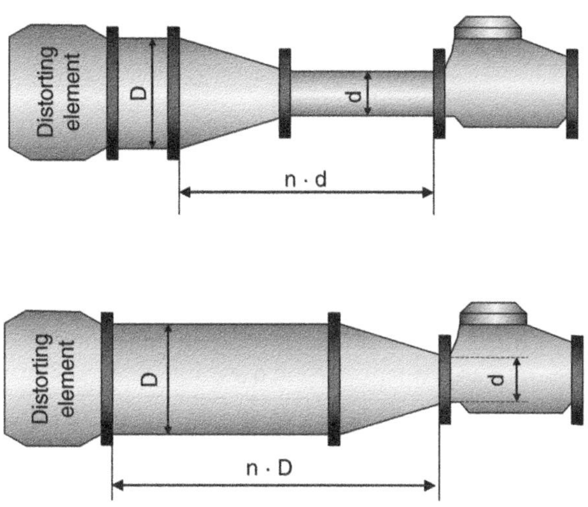

FIGURE 2.38. MEASUREMENT OF THE ADDITIONAL LENGTH OF PIPE REQUIRED WITH A DIAMETER REDUCTION

Where the available space is not enough to satisfy the recommendation from the manufacturer in terms of length of pipe upstream of the meter, flow stabilizers can be used. However, even in this case, is important to revise the recommendations of the manufacturer since depending on the model, the conditions for flow stabilizers can also be very different.

In some cases, the stabilizer can be installed immediately upstream from the meter (right part of Figure 2.37) while in others there must be a separation of at least one diameter from the meter. Additionally, and depending on the stabilizer, the need for additional segments of straight pipe will or will not be necessary. Last, but not least, it should always be taken into account that some flow distortions cannot be corrected by flow stabilizers (for instance the ones created by a sudden increase in diameter). Figure 2.39 shows the average influence of a gate valve installed three diameters upstream of a meter, at flowrates ranging from transitional flowrate to maximum flowrate. In the actual meter used (horizontal vane Woltmann, 80 mm) the metrology is not affected by the valve.

As explained before, the error curve of vertical vane Woltmann meters is by far less sensitive to the velocity profile. This is one important advantage of this configuration, for they require less room in the field for their installation. Usually, manufacturers recommend maintaining a distance of 3–5 diameters for a vertical vane Woltmann. Figure 2.40 shows the results of a test carried out for a vertical vane Woltmann to determine the effect of a partially closed gate valve and a partially closed butterfly valve three diameters upstream. The influence of these elements is almost negligible.

2.5.6. Metrology of Woltmann meters: Aging

Generally speaking, and taking into account only the aging of the meter, the error curve of a Woltmann meter should be displaced towards negative errors (underregistration).

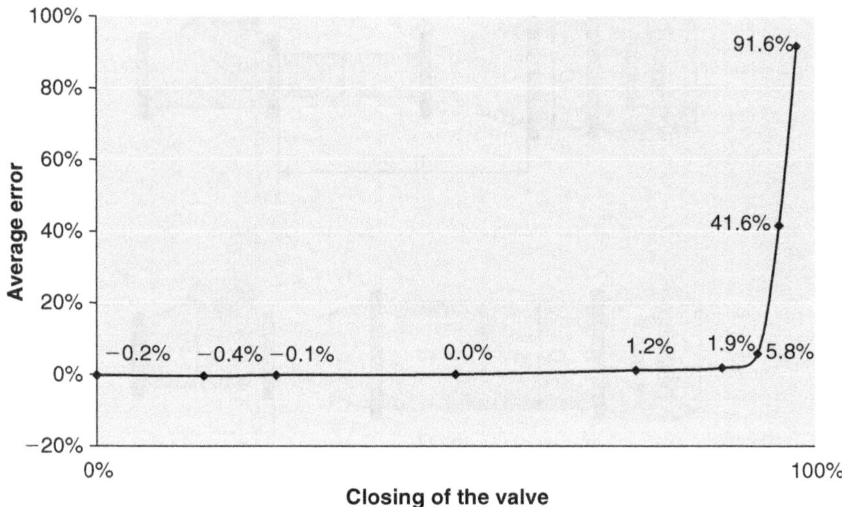

FIGURE 2.39. AVERAGE ERROR OF A HORIZONTAL VANE WOLTMANN WITH AN UPSTREAM GATE VALVE

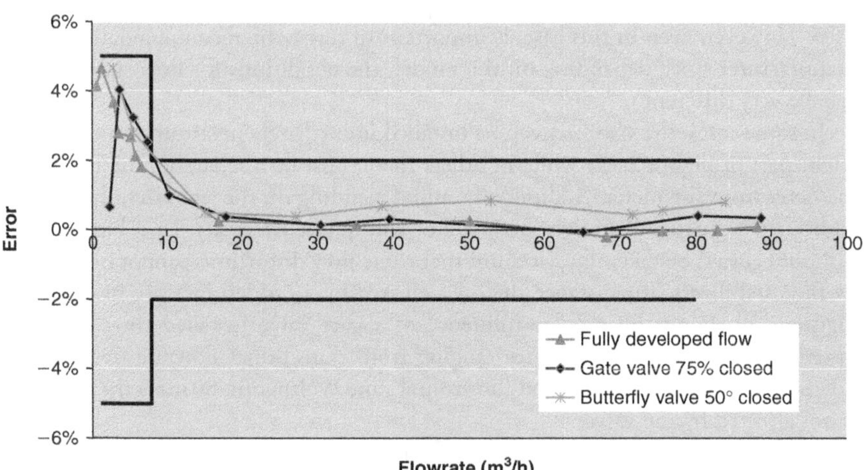

FIGURE 2.40. AVERAGE ERROR OF A VERTICAL VANE WOLTMANN WITH UPSTREAM GATE AND BUTTERFLY VALVES

The angular momentum that generates the motion in the turbine will not increase its value on the basis of aging. However, the wear of some of the meter elements will eventually lead to an increased resistance torque and negative errors.

The regulation mechanism and the large cross section prevent, most of the time, positive errors originated by solids build up inside the meter. Obviously, in an extreme situation the turbine could be blocked by depositions.

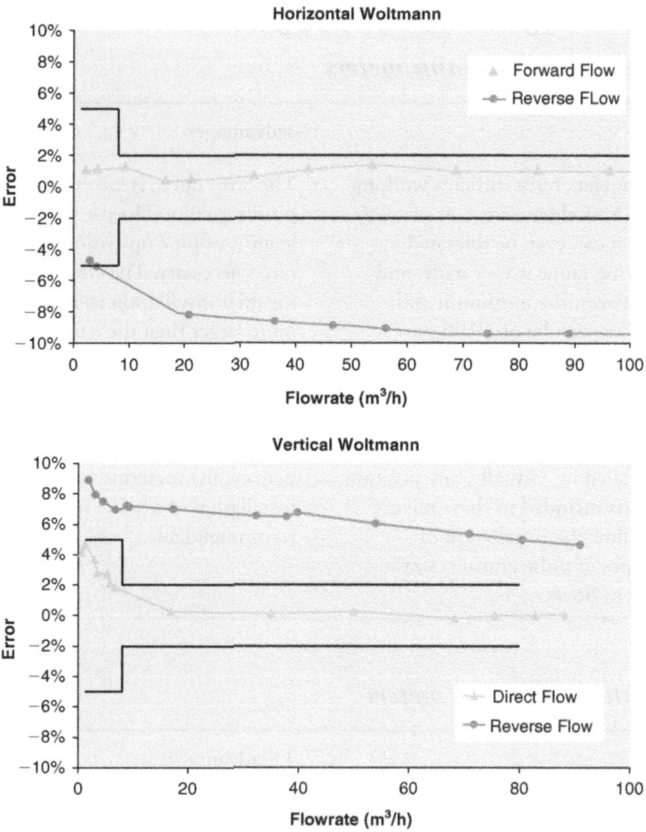

FIGURE 2.41. ERROR CURVE FOR BOTH FLOW DIRECTIONS IN VERTICAL AND HORIZONTAL VANE WOLTMANN METERS

2.5.7. Metrology of Woltmann meters: Reverse flow

Woltmann meters are not symmetrical, and it would seem logical that the metering errors would be different depending on the direction of the flow. Furthermore, meters usually have only one regulation plate located at the forward flow upstream side of the turbine. Consequently it only has an adjusting effect when water flows in forward direction.

Reverse accuracy is not a key issue for smaller meters but it becomes more important for larger meters installed in the network where it is difficult to guarantee that water will only flow in one direction.

The resulting errors can be quite important when, for instance, performing a water balance. However, there are horizontal vane Woltmann meters available in the market which according to the manufacturer guarantee similar errors for both directions of the flow.

Figure 2.41 shows the behaviour of two vertical and horizontal Woltmann meters, with water flowing in both directions. The two meters present significant differences in the error curve.

2.5.8. Advantages and disadvantages

Horizontal vane Woltmann meters

Advantages	Disadvantages
• Woltmann meters resist difficult working conditions. A moderate amount of solids in suspension can even be tolerated. • The measuring range is very wide, and the ratio between the minimum and maximum flow can be of 1:100 or even 1:200. • The metering module may be replaced without affecting the meter accuracy, keeping the meter housing. • Can be installed in, virtually, any position. • The totalizers included in these meters nowadays allow the installation of different types of pulse emitters so they can be used as flowmeters.	• The error curve is sensitive to the velocity profile quality. The use of tranquilizing lengths of pipe upstream of the meter is often necessary. The effective space needed for their installation is, as a consequence, quite larger than the length of the meter. • Low sensitivity for low flowrates. Currently, there are no Class C meters available in this technology. • The impact of large suspended solids could damage the metering module, making the installation of a stone strainer quite recommendable.

Vertical vane Woltmann meters

Advantages	Disadvantages
• Good behaviour at low flowrates. • Reduced sensitivity to the quality of the velocity profile, and almost no need for additional space.	• Expensive. • Less capacity than an equivalent diameter horizontal vane meter. • Fixed installation position.

2.6. COMBINATION (COMPOUND) METERS

2.6.1. Operating principles

The design of new meter technologies is usually not aimed to further reduce the measuring errors, but rather focuses on providing a wider range of measurable values. As a matter of fact, the quality of meters as understood under ISO 4064 is dependant on the measuring range and not on the error tolerances, which are identical in all metrological classes.

A technical solution to increase the effective range of measurement, designed a few decades ago, is the combination meter. This type of device incorporates two or three meters of several diameters and capacities in a single housing to provide a wider range. Each meter is used for certain flowrates, combining the metrological virtues of all of them. These meters are used in industrial applications, hospitals, schools, etc. where there are important differences between minimum and peak flows.

2.6.2. Constructive characteristics

With regards to the path followed by water, combination meters are arranged in parallel, as shown in Figure 2.42. The derivation of the flow through one circuit or the other is achieved by an automatic valve (changeover) installed downstream from the main meter. When the flowrate is low, the valve is closed and water only flows through the secondary meter. If the flowrate is increased, the pressure loss in the secondary meter is increased causing the crossover valve to open. When this happens water flows through both meters.

The meter which has a greater capacity is usually a Woltmann, although sometimes industrial volumetric meters are used. The secondary meter may also be a volumetric one, multiple or single jet meter.

For the construction of combination meters, manufacturers usually choose one of two alternatives:

- Integrate both meters in a single casing, providing a "series" appearance and saving some installation space, which can be quite useful (Figure 2.43, right).

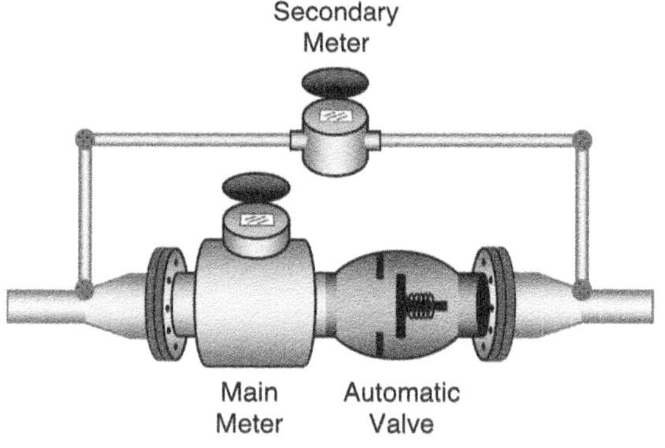

FIGURE 2.42.   INTERNAL SETUP FOR A COMBINATION METER

FIGURE 2.43.   PARALLEL AND SERIES CONFIGURATIONS FOR COMBINATION METERS
(PARALLEL COURTESY OF ACTARIS – SERIES COURTESY OF SENSUS)

- Maintain both meters in separate casings, with the corresponding ducts to communicate them. This construction accentuates the real parallel arrangement of the meters (Figure 2.43, left).

Figure 2.44 shows the cross section of a series combination meter, where the crossover valve can be seen in operation depending on the circulating flow. The figure

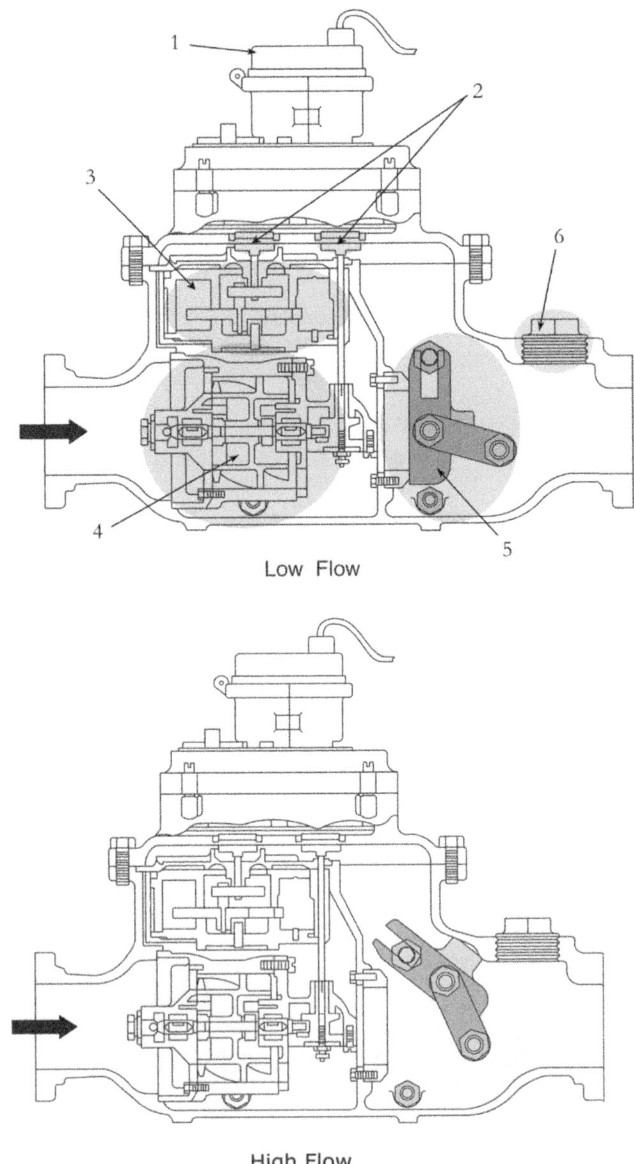

FIGURE 2.44. CROSS SECTION AND OPERATION OF A COMBINATION SERIES METER

also shows the components of a combination meter:

1. Totalizer: Usually mechanical. In the figure, there is only one totalizer, although more commonly each meter will have an independent totalizer and the readings from both must be added to get the total. This is quite inconvenient for many utilities, for it usually requires to include the possibility of having two meters associated to a single customer in the commercial database system.
2. Magnetic couplings.
3. Measuring chamber of the secondary meter. In this case, it is a rotating piston meter.
4. Measuring chamber of the main meter. Usually, a horizontal vane Woltmann meter.
5. Automatic changeover valve. This element needs frequent maintenance to ensure the correct operation of the meter.
6. Test connection to check the metering error.

2.6.3. Metrological characteristics and dimensions

Combination meters are used to register volumes of important customers with wide variations in consumption flowrates. They can usually be found in diameters ranging from 50 to 250 mm.

The final characteristics of the meter will depend on the technology and nominal flow of the meters that are finally combined, and whether the combination is in series or in parallel. Tables 2.14 and 2.15 show the characteristics of two types of combination meters: a vertical vane Woltmann with a rotating piston and a horizontal vane Woltmann with a multiple jet meter.

Neither table shows *official* metrological parameters (such as minimum, transitional and nominal flowrates). The reason is that it is very difficult to find certified meters using this technology. When available, they are Class B meters, which is inconsistent with their theoretical advantage in a wide range of flows since it should produce a much higher class. In any case, separately, the meters are usually ISO 4064 compliant.

TABLE 2.14. METROLOGICAL CHARACTERISTICS OF A COMBINATION VERTICAL VANE WOLTMANN – ROTATING PISTON METER (SERIES)

Vertical vane Woltmann (main)				
Diameter (mm)	50	80	100	150
Over flowrate (m^3/h)	45	147	262	568
Permanent flowrate (m^3/h)	36	73.5	131	355
1 bar losses flowrate (m^3/h)	42	107	163	430
Rotating piston (secondary)				
Diameter (mm)	15	20	20	30
Minimum flowrate (l/h)	15	25	25	60
Starting flowrate (l/h)	3.4	3.4	3.4	13.6
Crossover valve				
Changeover flowrate (rising) (m^3/h)	2	3	3.3	7.2
Changeover flowrate (falling) (m^3/h)	1	1.5	1.9	4.6

TABLE 2.15.   METROLOGICAL CHARACTERISTICS OF A COMBINATION
HORIZONTAL VANE WOLTMANN – MULTIJET METER (SERIES)

Vertical vane Woltmann (main)				
Diameter (mm)	50	65	80	100
Over flowrate (m^3/h)	90	90	200	250
Permanent flowrate (m^3/h)	50	50	120	180
1 bar losses flowrate (m^3/h)	47	47	103	161
Rotating piston (secondary)				
Diameter (mm)	20	20	20	20
Minimum flowrate (l/h)	20	20	20	20
Starting flowrate (l/h)	37.5	37.5	37.5	37.5
Crossover valve				
Changeover flowrate (rising) (m^3/h)	2	2	2.2	2.8
Changeover flowrate (falling) (m^3/h)	1	1	1.2	1.8

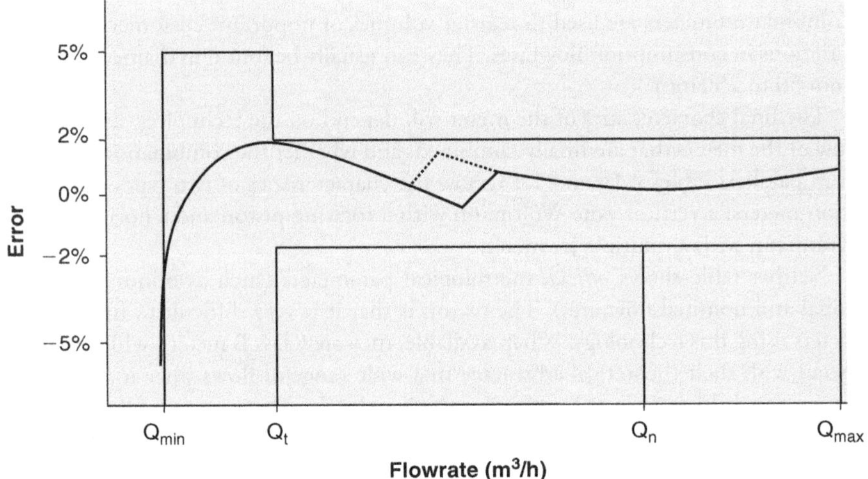

FIGURE 2.45.   TYPICAL ERROR CURVE OF A COMBINATION METER

The range of these meters, taking into account the data provided by the manufacturer, is extraordinary. For instance, the 100 mm meter from Table 2.15 presents a 1:4800 ratio between the permanent flowrate and the secondary transitional flowrate (this is the range where the combination meter presents errors lower than ±2%). This feature is the best characteristic of the combination meters.

The error and pressure losses curves of these meters are very distinctive and they show the moment that the main meter starts/stops functioning. This can be seen in Figure 2.45, which like the pressure loss curve (Figure 2.46) shows a small hysteresis cycle.

Figure 2.46 shows the pressure losses for a series combination meter. In the graph, starting at 10 m^3/h three different curves can be observed, corresponding to the three different diameters in which the combination meter is sold. Additionally, it is also possible to identify a hysteresis cycle in the operation of the changeover valve.

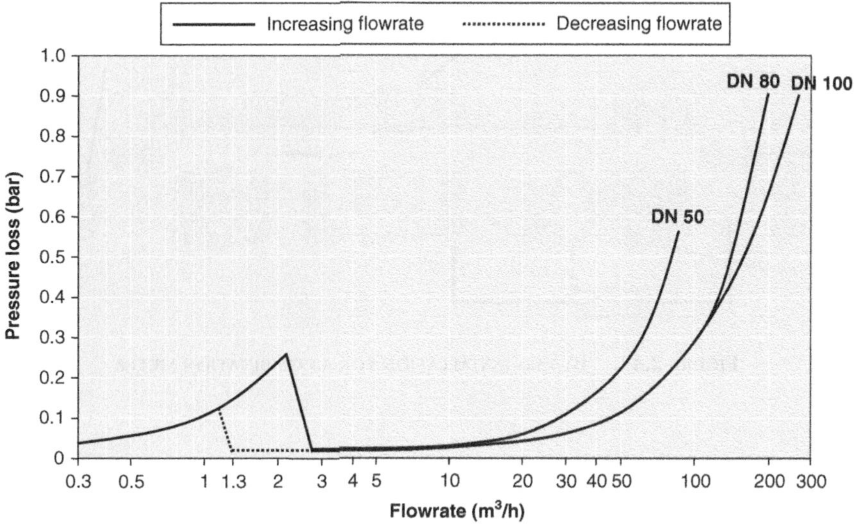

FIGURE 2.46. TYPICAL PRESSURE LOSS CURVE OF A COMBINATION METER

2.6.4. Metrology of combination meters: Orientation

Combination meters are made up of two different meters. As a consequence, the correct orientation for the combination depends on the characteristics of both meters. In practice, the main meter is the one imposing most restrictions for the installation, for it is usually a Woltmann meter. Depending on the type of secondary meter, the orientation restrictions will be more or less demanding.

If the main meter is a horizontal vane Woltmann, it can be installed in non-horizontal and even vertical pipes, with even some rotation around the pipe axis. However, in the case of vertical vane Woltmann meters, the only possible orientation is horizontal with the totalizer in vertical position. The combination parallel meters also demand this sort of installation.

2.6.5. Metrology of combination meters: Velocity profile

The requirements for the quality of the velocity profile in combination meters are the same ones as for the Woltmann meters. Some manufacturers simplify the problem and only request 5 or 10 diameters of straight pipe independently of the type of perturbation, which in any case should be studied and stabilized.

It is recommendable to never install a regulation or distorting element upstream from the meter. Another possibility to be taken into account is to provide an alternative circuit in parallel. This configuration allows to carry out the frequent maintenance tasks that are necessary without having to interrupt the supply (Figure 2.47).

Finally, it is important to point out that many manufacturers recommend the use of strainers to protect the meter (impeller) from the possible impact of suspended solids. American manufacturers recommend flat or vertical Z-plate strainers (recommended by the AWWA), like the one shown in Figure 2.48. These strainers not only prevent the damages caused by solids, but also regularize the velocity profile improving the measure.

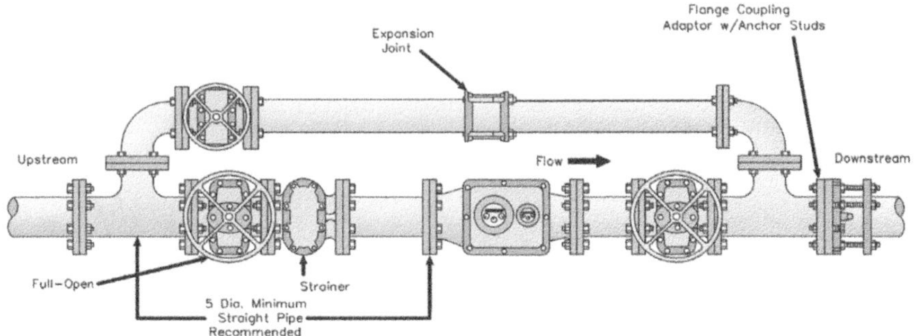

FIGURE 2.47.   BY-PASS INSTALLATION FOR A COMBINATION METER

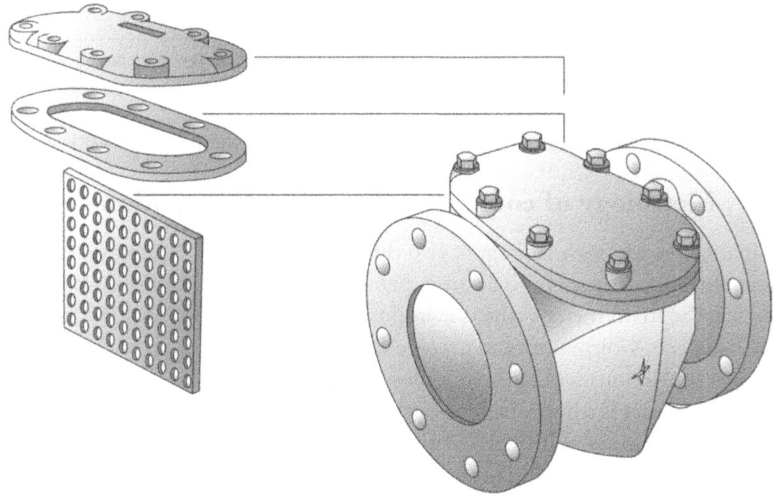

FIGURE 2.48.   AWWA RECOMMENDED STRAINER FOR COMBINATION METERS

Distinction should be made between this strainer which may even help to reduce distortions in the velocity profile and the typical ring or mesh strainers which are used, for instance, in irrigation. These strainers introduce important perturbation in the flow and have to be installed with a certain length of straight pipe between strainer and meter.

2.6.6.   Metrology of combination meters: Aging

Just like with all metering technologies, wear of the mobile parts in combination meters leads to increased friction and difficulties in the motion, and as a consequence, to undermetering. This effect is more noticeable at lower flowrates.

It is relatively frequent to find the secondary circuit to be completely blocked. In combination parallel meters, this means that the main meter will not measure anything until the crossover flow is reached. When the meters are placed in series, the measure of the main meter is conditioned by its metrology and not the crossover flow, which greatly reduces the chances of a significant error.

Combination meters are not being used any more in many countries. The reason for this is the need to maintain the changeover valve. The fact is that the quality of the measurement is not only dependant on the mobile components to be in perfect shape, but also on the adequate behaviour of the mentioned valve.

The main problem appears when the valve does not close the main circuit completely at low flowrates, producing leaks through this circuit. This flowrate is not large enough to move the impeller within its effective range of operation and it fails to register partially or totally such volume.

2.6.7. Metrology of combination meters: Maintenance

As a consequence of the previous point, it is essential to carry out a preventive maintenance programme of combination meters, including revisions and cleaning operations. The revision of the meters should include missing pieces, loose screws, marks in the totalizer's display and, in general, any other sign of unauthorized manipulation of the meter. Additionally, it is convenient to verify that the meter is operating correctly within an adequate range of flowrates, and the pressure losses are not excessive. If the pressure losses were too large this would be a possible sign of the circuit being blocked or some of the mobile components being broken.

The periodical cleaning of the meters should be done for both the housing and the interior of the meter. The later would ensure that the changeover valve operates correctly and that the main impeller or the secondary circuit is not blocked.

Additionally, and when necessary, there are available models in the market in which it is possible to replace the metering module without affecting the metrology of the meter.

2.6.8. Advantages and disadvantages of combination meters

Advantages	Disadvantages
• Very wide measuring range.	• Need for maintenance. • High probability of failures in the changeover valve leading to important metering errors. • Many models use separate totalizers for the main and the secondary meters. • Expensive. • Not suitable for "dirty" waters, especially those carrying suspended solids.

2.7. PADDLE WHEEL METERS

2.7.1. Operating principles

Paddle wheel meters, often referred to by the manufacturers as *irrigation meters*, are very similar to some types of insertion flow meters. The flow is measured by an impeller placed on the top section of the meter (paddle wheel). This method also has similarities with single jet meters where water hits only part of the impeller and the rotation is perpendicular to the pipe axis.

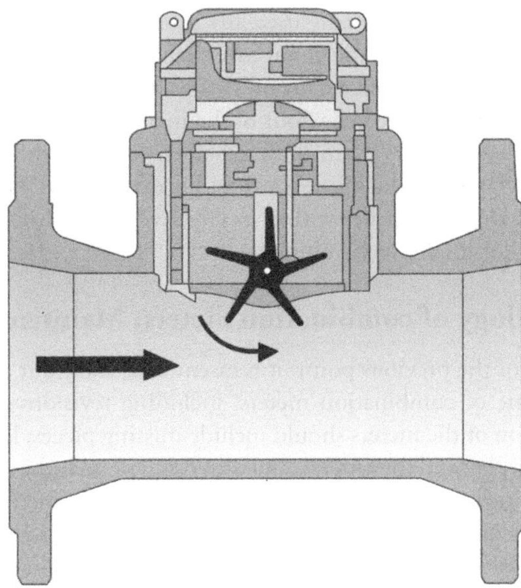

FIGURE 2.49. CROSS SECTION OF A PADDLE WHEEL OR IRRIGATION METER

Figure 2.49 shows a longitudinal section of a paddle wheel meter. The rotation velocity of the turbine is proportional to the velocity of water in the upper section of the meter. It is important to note that the meter is not integrating the velocity in the whole section of the meter, but doing it on a very limited section instead. Obviously, in this situation, an asymmetric velocity profile or with radial components will result in differences between the average velocity in the pipe and the meter's registered velocity, seriously affecting accuracy.

2.7.2. Constructive characteristics

Figure 2.50 shows the interior of a paddle wheel meter. The housing has room for the impeller in the upper part, while the inner diameter remains constant for the whole meter. In some models, the metering section is shaped like a Venturi (gradual diameter reduction). This shape presents the advantage of regularizing the velocity profile, slightly reducing the sensitivity of the measure to upstream perturbations of the flow.

The interior of the meter is often coated with epoxy (most large meters are) in order to improve the resistance to corrosion, especially important with aggressive water.

In the upper part of the section, a deflector vane orientates the water flow towards the impeller. This device achieves an angle near to 90° between water and impeller.

The rotation of the impeller, perpendicular to the gears of the totalizer, requires an endless screw or a bevel gear. The result, as with other technologies, is that the sensitivity of the meter to low flows is reduced.

FIGURE 2.50. INTERIOR OF A PADDLE WHEEL METER

2.7.3. Metrological characteristics and dimensions

The measuring performance of paddle wheel meters is significantly lower than that achieved by other technologies. However, the great advantage of these meters is that most of the cross section is free, allowing suspended solids, even small stones, to flow through the meter without damage. This is especially relevant in the case of irrigation water, one of the great applications of this technology.

With regards to metrology it is important to point out that even in 2000 it was difficult to find paddle wheel meters complying with ISO 4064:1993. From that year onwards the number of Class A models increased significantly, and a few models even reached Class B.

The difference between the meters reflected in Tables 2.16 and 2.17 lies in their capacity to maintain a high accuracy at low flows. The Class A meter shown in Table 2.16 has been designed to provide optimum performance both at high and low flowrates.

Paddle wheel meters are a good option regardless of their narrow range of measure and their high sensitivity to the velocity profile. The main reason is that they are reasonably priced, for the metering device maintains its size independently of the diameter of the meter.

2.7.4. Metrology of paddle wheel meters: Orientation

If the meter is installed in an inclined position, the advantage just described concerning suspended solids is lost. Additionally the impeller shaft does not rest adequately and is subject to greater wear, reducing the sensitivity at low flows, which is already quite low.

TABLE 2.16. METROLOGICAL CHARACTERISTICS OF A CLASS B PADDLE WHEEL METER

Diameter (mm)	50	65	80	100	125	150	200
Nominal flowrate (m³/h)	50	60	75	150	175	250	400
Maximum flowrate (m³/h)	100	120	150	300	350	500	800
Transitional flowrate (m³/h)	1.5	1.8	2.25	4.5	5.25	7.5	12
Minimum flowrate (m³/h)	1	1.2	1.5	3	3.5	5	8
Service pressure (bar)	16	16	16	16	16	16	16
Pressure loss at maximum flowrate (bar)	0.1	0.1	0.1	0.1	0.1	0.1	0.1

TABLE 2.17. METROLOGICAL CHARACTERISTICS OF A CLASS A PADDLE WHEEL METER

Diameter (mm)	50	65	80	100	150
Nominal flowrate (m³/h)	15	25	40	60	150
Peak flowrate (m³/h)	60	100	150	250	500
Permanent flowrate (m³/h)	30	50	90	140	250
Transitional flowrate (m³/h)	5	7	10	15	35
Minimum flowrate (m³/h)	1	2	3	5	18
Service pressure (bar)	16	16	16	16	16
Pressure loss at maximum flowrate (bar)	<0.2	<0.2	<0.2	<0.2	<0.2

2.7.5.　Metrology of paddle wheel meters: Velocity profile

As described previously, a paddle wheel meter only integrates the flow in the higher part of the cross section of the meter. Any distortion of the velocity profile will consequently affect the error. As a general rule, the larger the meter, the smaller the section considered for metering. Consequently, large meters are in principle more sensitive to perturbations of the velocity profile.

It is then essential to guarantee that paddle wheel meters are installed in such a way that the velocity profile has an excellent quality, and so the needed lengths of straight sections of pipe upstream from the meter are considerable. As a reference, the recommendations made by manufacturers of turbine insertion flow meters could be used (15–30 diameters of straight pipe upstream from the meter, depending on the perturbation). However, some meter manufacturers only require five diameters. This distance may be insufficient, as some tests carried out by the authors at the Instituto Tecnológico del Agua show. These tests evaluated the influence of two elbows out of plane and a butterfly valve placed upstream from the meter. Figure 2.51 shows that even at a distance of seven diameters, the error is still above the required ±2%.

Another example can be found when studying the effects of a partially closed gate valve at a distance of three diameters. The effects are surprisingly inaccurate. Even with a completely open valve, the underregistration can reach −10%. What really happens in this case, is that the velocity in the lower part of the section is increased considerably when compared with the velocity in the upper part of the section (which is the one being registered by the meter). As a consequence, the meter "believes" that the flowrate is much lower than it really is (Figure 2.52).

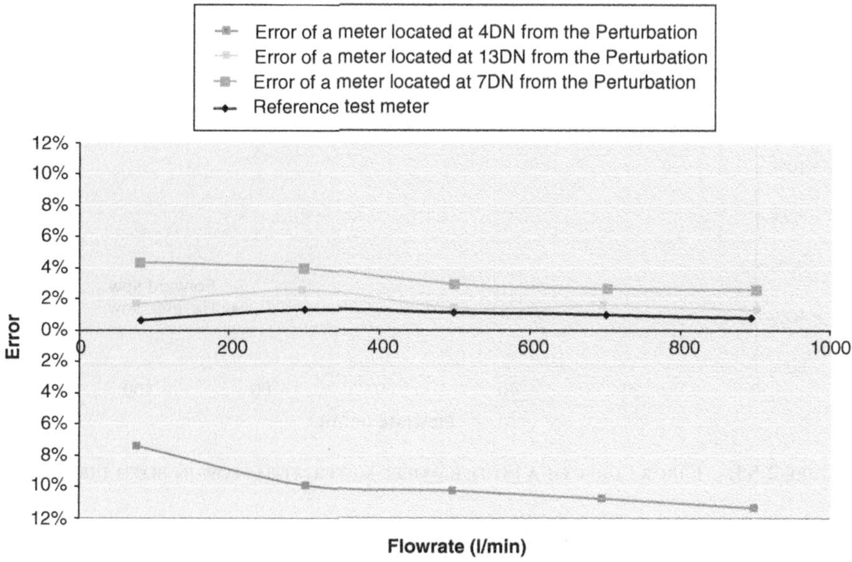

FIGURE 2.51. ERROR CURVE OF A PADDLE WHEEL METER WITH A DISTORTION OF THE VELOCITY PROFILE PLACED AT SEVERAL DISTANCES

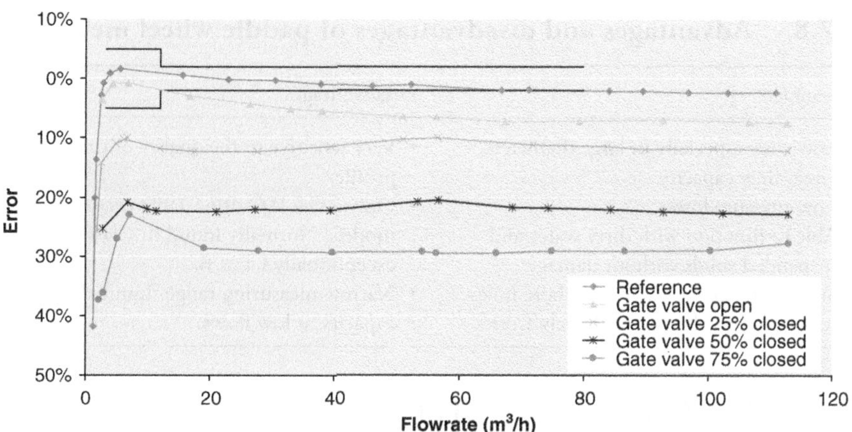

FIGURE 2.52. ERROR CURVE OF A PADDLE WHEEL METER WITH A GATE VALVE PLACED AT A DISTANCE OF 3D UPSTREAM

2.7.6. Metrology of paddle wheel meters: Reverse flow

The metering error in a paddle wheel meter is also affected by the direction of the flow. The flow deflectors placed next to the impeller are not completely symmetrical (the one upstream is different from the one downstream) (Figure 2.53).

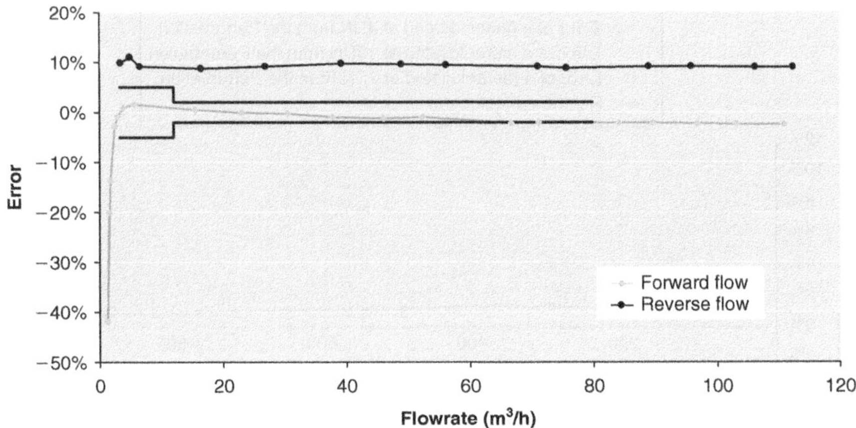

FIGURE 2.53.   ERROR CURVE OF A PADDLE WHEEL METER WITH FLOW IN BOTH DIRECTIONS

2.7.7.   Metrology of paddle wheel meters: Suspended solids

The main advantage of paddle wheel meters is that water finds no obstacles in flowing through the meter, allowing large suspended solids to go through the meter without damaging the impeller. Higher-density solids, which could damage the impeller, will usually travel in the lower part of the pipe while the impeller is situated in the higher part.

2.7.8.   Advantages and disadvantages of paddle wheel meters

Advantages	Disadvantages
• Low cost, especially in large diameters. • Great flow capacity. • Low-pressure losses. • Able to function with dirty water and suspended solids without damage. Suitable to meter water from bore holes (cross section almost completely free).	• Very sensitive to the quality of the velocity profile. • Only a few ISO 4064:1993 certified models. Normally found in Class A, and exceptionally Class B. • Narrow measuring range. Limited metering capacity at low flows.

2.8.   PROPORTIONAL METERS

2.8.1.   Operating principles

Proportional meters are based in the relationship between the flows circulating through two parallel circuits (Figure 2.54) and the total flowrate of the pipe.

Pressure losses through both circuits (between points 1 and 2) must necessarily be the same. Additionally the sum of the two partial flowrates must equal the total flowrate:

$$\left.\begin{array}{l} h_{1-2}^{\text{main}} = K_{\text{main}} \cdot Q_1^2 \\ h_{1-2}^{\text{secondary}} = K_{\text{secondary}} \cdot Q_2^2 \end{array}\right\} \rightarrow h_{1-2}^{\text{main}} = h_{1-2}^{\text{secondary}} \qquad (2.1)$$

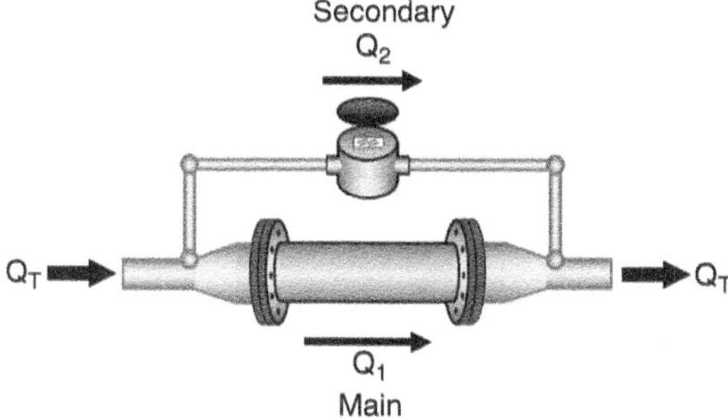

FIGURE 2.54. RELATIONSHIP BETWEEN PARALLEL CIRCULATING FLOWS

$$Q_T = Q_1 + Q_2 \tag{2.2}$$

$$Q_1 = \sqrt{\frac{K_{secondary}}{K_{main}}} \cdot Q_2, \text{ hence } Q_T = \left(\sqrt{\frac{K_{secondary}}{K_{main}}} + 1 \right) \cdot Q_2 \tag{2.3}$$

If the hydraulic resistance of each circuit is known (K_{main} and $K_{secondary}$) it is thus possible to obtain the total flowrate by measuring the flowrate in only one of the circuits (in this case the secondary one).

2.8.2. Constructive characteristics

Proportional meters use the presented relationship to control the circulating flowrate through a pipe. The device presents two parallel circuits with different capacities. The secondary circuit (of a smaller capacity) presents a single or multiple jet meter, while the main circuit presents a nozzle or diaphragm that allows water and even suspended solids to flow through, but at the same time generates a difference in pressure between points 1 and 2. This allows the flow that circulates through the secondary circuit to be correctly measured by a Class B meter and conveniently converted to the total flow through the meter. Figure 2.55 shows a proportional meter and its schematic.

Figure 2.56 shows the different parts of a proportional meter. The body of the meter hosts the main circuit with its nozzle or diaphragm. It must be taken into account that these elements enable the amount of water flowing through the secondary circuit to be significant. Otherwise, the meters to be used in this circuit should be very small or very sensitive, and would be affected by the quality of water.

The strainer is used to avoid the circulation of particles through the secondary circuits. This element is potentially one of the greatest sources of problems in proportional meters and requires adequate maintenance. Should the strainer get clogged, the resistant characteristics of both circuits might even change and the error curve would be seriously affected towards underregistration.

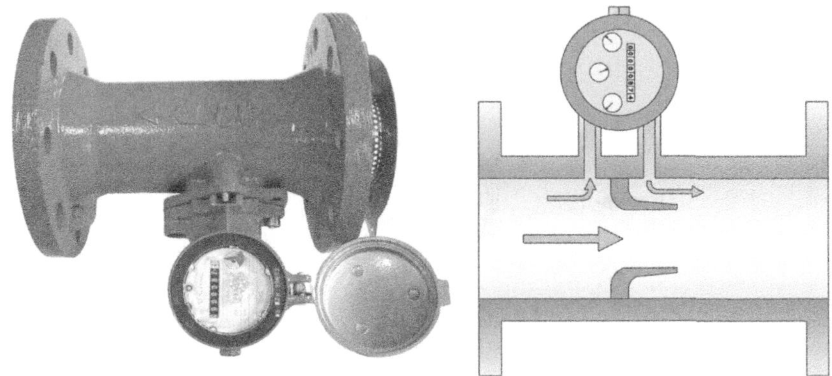

FIGURE 2.55.   PROPORTIONAL METER CONSTRUCTIVE CHARACTERISTICS

FIGURE 2.56.   PARTS OF A PROPORTIONAL METER

TABLE 2.18.   METROLOGICAL CHARACTERISTICS OF A PROPORTIONAL METER

	65	80	100	150
Diameter (mm)	65	80	100	150
Exceptional flowrate (m^3/h)	100	160	250	550
Permanent flowrate (m^3/h)	80	120	200	500
Flowrate at $\pm 5\%$ accuracy (m^3/h)	2.5	4.8	8	20
Starting flowrate (m^3/h)	0.8	1.8	3	7.5
Pressure loss at maximum flowrate (bar)	0.5	0.44	0.58	0.75
Length (mm)	255	255	255	255

2.8.3.   Metrological characteristics and dimensions

It is important to take into account that this type of meter is not usually certified in any metrological class. As a consequence, it usually cannot be used as a metering device for billing purposes.

Table 2.18 shows the main metrological characteristics of a proportional meter provided by a manufacturer. It can be seen that the sensitivity is quite poor. However, these meters present a great capacity, even when compared to horizontal vane Woltmann equivalents.

Proportional meters are mainly used to measure flows where the water has a lot of suspended solids or fibres. Their price is quite low, and for this reason they are also used in facilities where the flowrate does not change too much and velocities are medium to high (such as pumping stations, bore holes or localized irrigation systems). In any case, it must always be taken into account that the quality of the measurements obtained with these instruments is much lower than with other technologies.

2.8.4. Metrology of proportional meters: Orientation

Proportional meters can be installed vertically, horizontally and even in inclined position without much alteration of the error curve. However, the metrological behaviour of the meter in the secondary circuit will be affected by the installation position, leading sometimes to a loss of sensitivity at low flowrates.

2.8.5. Metrology of proportional meters: Velocity profile

Proportional meters are not especially sensitive to variations in the velocity profile. However, some elements such as regulation valves or pumps may affect the quality of the measurement. Manufacturers, recommend installing sections of straight pipe ahead of the meter (i.e. five diameters of length after a valve and 20 after a pump).

2.8.6. Metrology of proportional meters: Aging

The volume in proportional meters is measured through a meter placed in the secondary circuit. This is usually a single or multiple jet meter. Taking this fact into account, all considerations related to aging and wear for these technologies should be considered for proportional meters, including a maintenance plan.

2.8.7. Metrology of proportional meters: Reverse flow

The water flow through the secondary circuit is created through a difference in pressures between the two common points on the main circuit. If this pressure difference is created by an increment in the kinetic energy of the water (Figure 2.55) a reverse flow will not produce such effect (pressure difference will depend on pressure losses and differences in the kinetic energy) and the circulating flowrate through the secondary will certainly be smaller than the one with forward flow conditions. Figure 2.57 shows the results of a series of tests performed on a proportional meter, where the under-registration with reverse flow may reach up to -40%.

2.8.8. Metrology of proportional meters: Strainer clogging

The partial clogging of the strainer in proportional meters may lead to changes in the relationship between the resistant characteristics of the main and secondary circuits, and as a consequence, of the distribution in the circulating flowrates through both of them. This latter change is the one that affects the most the error curve of the meter. It is important that the meter is designed to prevent the strainer clogging.

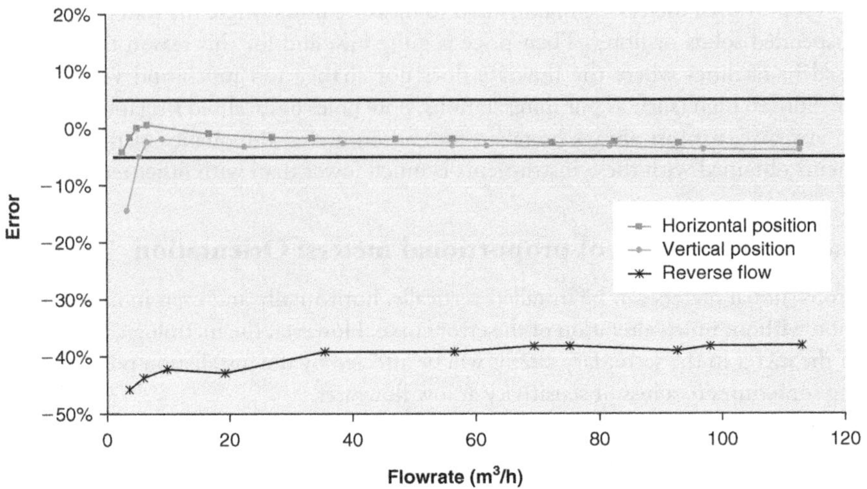

FIGURE 2.57. PROPORTIONAL METER BEHAVIOUR WITH REVERSE FLOW

2.8.9. Advantages and disadvantages of proportional meters

Advantages	Disadvantages
• Low cost. • Great flow capacity. • Low-pressure losses. • Suitable for "dirty" water. May be used in bore holes.	• Unreliable technology. • There are no ISO4064 certified models. • Small measuring range. Very limited at low flows. • Limited accuracy. Manufacturers usually only guarantee errors within ±5%.

3

Flow meters

3.1. ULTRASONIC FLOW METERS

3.1.1. Operating principles

The ultrasonic flow meter is one of the metering technologies with the greater impact in the past few years. For large diameters it is the most cost effective solution, and sometimes the only one available. This technology is very versatile and can be used in diameters of up to 8000 mm.

There are basically two types of ultrasonic flow meters that respond to physical principles totally different in nature. The first one uses the variation of the absolute velocity of sound travelling in a moving media (Figure 3.1). The second one uses the change in frequency of a sound wave when it is reflected in a body at motion when observed from a fixed position (the well-known Doppler effect). The first type of flow meters are known as transit time ultrasonic flow meters, while the others are known as Doppler effect ultrasonic flow meters.

3.1.2. Transit time ultrasonic meters

Transit time meters take advantage of the fact that sound travels at a different velocity depending on the medium it uses to travel. It is of common knowledge that sound travels faster in water than through the air, and these differences can be found in any medium depending on its physical characteristics. When these media are also in motion with respect to an observer the velocity of sound relative to that observer changes as

FIGURE 3.1. ULTRASONIC METER IN A LARGE-DIAMETER PIPE (COURTESY OF RITTMEYER)

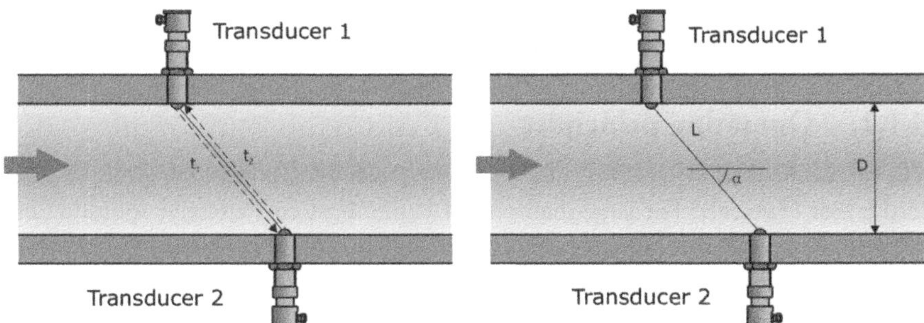

FIGURE 3.2. OPERATING PRINCIPLE OF A TRANSIT TIME ULTRASONIC FLOW METER

well (sound travelling through air with or without wind). Transit time ultrasonic meters are based on this property of sound (Figure 3.2).

As a general rule, an ultrasonic flow meter uses two ceramic piezoelectric transducers that can work both as emitters and receivers of sound waves. When acting as emitters they are excited by series of electric pulses that are converted to acoustic signals. When acting as receivers the process is inverted, and the pressure waves are converted into electric pulses.

Transducers can be classified into wetted transducers (in contact with the fluid) and clamp-on transducers, which are installed externally. The first ones provide a more accurate measure, while the second ones are much more versatile from an installation point of view.

Sensors are usually placed separated by a distance L and in opposite diameters. The upstream transducer emits a sound signal which is picked up by the downstream receiver after a time t_1. Then the process is reversed and the downstream transducer acts as the emitter where t_2 is the time the sound takes to travel from the downstream to the upstream transducer.

If the velocity of the fluid in the pipe is zero, both times are identical. However, when the fluid flows with a certain velocity, the transmission velocity of the sound changes (it is increased in the direction of the flow and reduced in the opposite direction). Therefore, the velocity of the sound in that medium c is modified by the velocity of the flow. The difference between the times $t_2 - t_1$ is the parameter actually used to calculate the average flow velocity and the flowrate. Mathematically:

$$t_1 = \frac{L}{c + V \cdot \cos \alpha} \tag{3.1}$$

$$t_2 = \frac{L}{c - V \cdot \cos \alpha} \tag{3.2}$$

And hence:

$$t_2 - t_1 = \Delta t = \frac{2 \cdot L \cdot V \cdot \cos \alpha}{c^2 - V^2 \cdot \cos^2 \alpha} \tag{3.3}$$

Taking into account that the speed of sound in the medium is considerably higher that the velocity of the flow, and that $\cos^2 \alpha$ is always less than 1, it is possible to neglect the term $V^2 \cdot \cos^2 \alpha$ with respect to c^2. This enables the simplification of the equation providing the transit times to:

$$\Delta t = \frac{2 \cdot L \cdot V \cdot \cos \alpha}{c^2} \tag{3.4}$$

The average velocity of the fluid at the cord between the transducers is then:

$$V = \frac{\Delta t \cdot c^2}{2 \cdot L \cdot \cos \alpha} \tag{3.5}$$

The distance between the transducers, L, is related to the diameter by the following expression:

$$L = \frac{D}{\sin \alpha} \tag{3.6}$$

Finally obtaining:

$$V = \frac{\Delta t \cdot c^2 \cdot \sin \alpha}{2 \cdot D \cdot \cos \alpha} = \frac{\Delta t \cdot c^2 \cdot \tan \alpha}{2 \cdot D} \tag{3.7}$$

This last expression relates the difference in times, Δt, to the speed of sound in the medium c, the diameter of the pipe, the angle α and the average velocity of the fluid in the plane where the transducers are installed.

Additionally, the speed of sound in the medium can be obtained from the total time that the wave spends in travelling both ways:

$$c = \frac{2 \cdot L}{t_1 + t_2} = \frac{2 \cdot D}{\sin \alpha \cdot (t_1 + t_2)} \tag{3.8}$$

which combined with the previous expression allows determining the velocity of the fluid independently of the speed of sound in the medium:

$$V = \frac{\Delta t \cdot D}{\sin(2\alpha) \cdot (t_1 \cdot t_2)}$$

(3.9)

$$Q = A \cdot V = \frac{\pi \cdot D^2}{4} \frac{\Delta t \cdot D}{\sin(2\alpha) \cdot (t_1 \cdot t_2)} = \frac{\Delta t \cdot D^3 \cdot \pi}{4 \cdot \sin(2\alpha) \cdot (t_1 \cdot t_2)}$$

(3.10)

In any case, the speed of sound remains constant as long as the physical variables of the fluid (temperature, pressure, composition) remain constant. Consequently the velocity of the fluid will be obtainable simply from the difference in transit times and the known layout of the transducers.

3.1.3. Doppler effect ultrasonic flow meters

The operating principle of Doppler flow meters is based on the principle that the frequency of a sound wave changes if emitter and receiver are in relative motion. This same principle also applies when the waves are reflected on a body in motion (for instance the suspended particles in water). The frequency shift is proportional to the relative velocity of the particles and the emitter/receiver, and consequently to the flowrate.

Figure 3.3 shows the operating principle of Doppler effect meters with two external transducers. The emitter generates a series of sound pulses with a frequency f_i, that once reflected in the particles transported by water (suspended solids, gas bubbles) experiments a frequency shift of Δf. These variables are related by the following expression (c is the speed of sound in the medium and V is the average velocity of the particles present in the fluid flow):

$$\Delta f = 2 \cdot f_i \frac{V}{c}$$

(3.11)

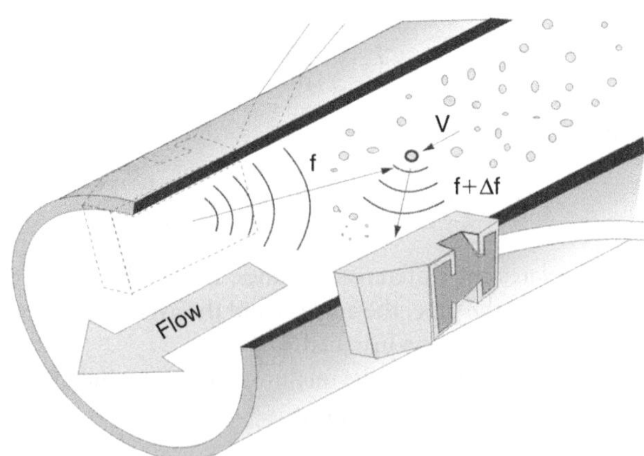

FIGURE 3.3.   OPERATING PRINCIPLE OF A DOPPLER EFFECT FLOW METER

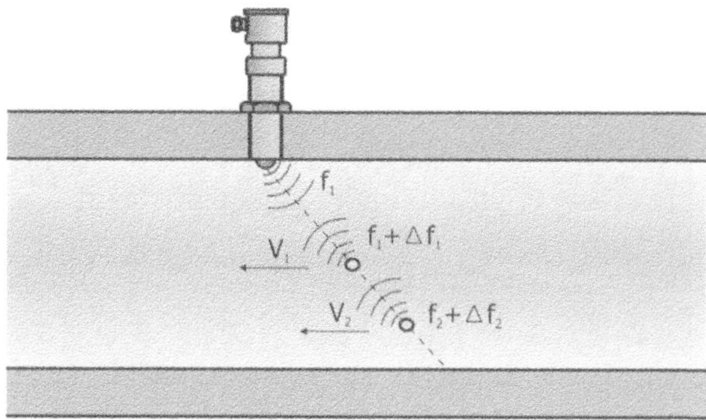

FIGURE 3.4. SETUP TO DETERMINE THE VELOCITY PROFILE BY MEANS OF A
DOPPLER EFFECT FLOW METER

The adequate operation of a Doppler meter requires the water to carry a minimum amount of suspended solids or gas bubbles to reflect the waves. The presence of these elements, however, is quite inconvenient for transit time meters, making both technologies complementary.

In some case, the Doppler effect can be used to determine the velocity profile in a pipe. Figure 3.4 shows a possible configuration. A transducer emits the sound waves, which when reflected in particles experience a frequency shift depending on their velocity. The waves reflected in the particles closer to the transducer will return sooner than the ones reflected in the particles which are further away. By measuring the frequency shift and the reception time it is possible to deduce the velocity profile and consequently improve the estimation of the flowrate.

In any case, it is important to keep in mind that these instruments do not measure the velocity of water directly but rather the velocity of the particles transported by water. After this, and when we know the cross section of the pipe, the flowrate is obtained. As a consequence these meters are much less accurate than transit time flow meters. The reason for this can be found, for instance, in a difference between the velocity of the particles and the fluid flow.

3.1.4. Constructive characteristics

As a general rule valid for all types, ultrasonic meters are composed of the transducers and the electronic components to create and process the signal.

A second classification of these meters could also be made depending on whether the transducers are wetted or clamped on the pipe. Figure 3.5 shows a transit time flow meter with clamp-on transducers. Doppler effect meters rarely use wetted transducers, favouring the clamp-on type with the exception of the models designed to measure open channel flow (Figure 3.7).

Constructive characteristics: Transit time meters

Besides the common characteristics described for both technologies, there are certain particular issues that only apply to transit time meters. As mentioned before, transit

FIGURE 3.5.	**PORTABLE ULTRASONIC METER**

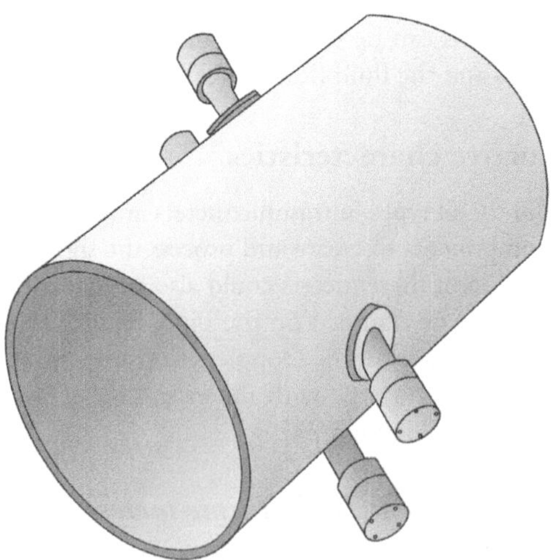

FIGURE 3.6.	**ULTRASONIC FLOW METER WITH WETTED TRANSDUCERS**

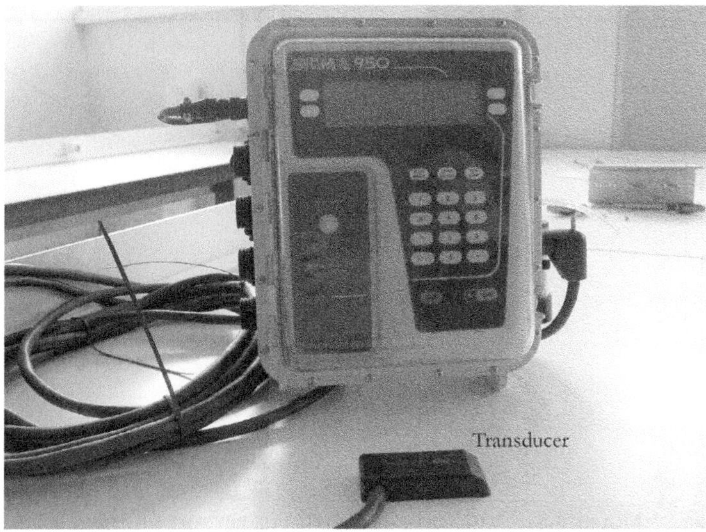

FIGURE 3.7. DOPPLER ULTRASONIC FLOW METER

time flow meters estimate the average velocity in the plane defined by the two transducers. The average velocity obtained is the average velocity in the path of the sound wave (V) which is not exactly the same as the average velocity of the fluid for the whole section $\bar{V}$. Should the relationship between these two velocities change, the flow measurement would be seriously affected.

Nikuradse (1932) carried out several tests with different flow conditions and obtained an expression relating both velocities by a proportionality factor k. The value of k depends on the type of section and flow. For fully developed velocity profiles in smooth wall circular pipes, the following expression relates the average velocity in the plane containing a diameter and the average velocity for the whole section: $\bar{V} = k \cdot V$, where:

$$k = \frac{1}{1.125 - 0.011 \cdot \log \mathrm{Re}} \tag{3.12}$$

In laminar flow ($\mathrm{Re} < 2000$) the value for k is 0.75 (for the plane containing the diameter). When Reynolds number reaches 10^5, k is 0.935. Consequently a proper measurement of the flow requires determining the Reynolds number. As a consequence it is necessary to focus not only on the velocity of the fluid, but also on possible changes of the viscosity and the pipe diameter (most ultrasonic meters nowadays allow to adjust these settings, including temperature – which is responsible for significant changes in viscosity).

One of the possible solutions to improve the accuracy in the flow estimation is to take several measurements in planes not containing the pipe axis. This is generally known as multi-path configuration, a setup that increases the number of transducers used. The objective is to have information about the flow velocity at several trajectories

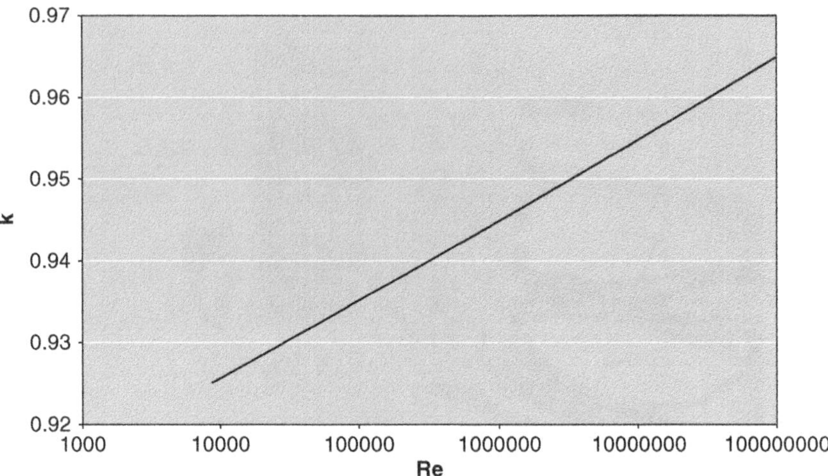

FIGURE 3.8. VARIATION OF *k* WITH THE REYNOLDS NUMBER

in the pipe in order to reduce the dependence of the measurements to possible velocity profile distortions. Figure 3.9 shows several of these configurations, including the Gauss-Chebyshev and the average radius. The first one is based in setting up a decagon with vertices 5 and 10 on the vertical diameter of the pipe, and joining vertices 1–9, 2–8, 3–7 and 4–6 creating four parallel planes. The average radius configuration places the planes approximately at a distance $R/2$ (precisely 0.52 R) from the pipe axis. The reason is that the relationship between the average velocity in such plane and the average velocity for the whole section is 1:1 (smooth turbulent flow) and remains reasonably constant (less than 0.5% in a range of Reynolds numbers).

The determination of the flowrate is done by weighting the average velocities in the different planes. Table 3.1 shows the different coefficients to be applied for each trajectory:

$$Q = A \cdot V = \frac{\pi \cdot D^2}{4} \sum w_i V_i \tag{3.13}$$

3.1.5. Metrologic characteristics and dimensions

Transit time ultrasonic flow meters

Transit time flow meters can be very accurate in certain configurations, reaching uncertainties as low as 0.2% of rate. However, it is important to distinguish between the accuracy obtainable for each setup.

Flow meters with wetted transducers are very reliable and achieve accuracies between 1% and 0.25%. As a rule of thumb, the higher the number of transducers and trajectories, the lower the uncertainty and the smaller the sensitivity to the quality of the velocity profile.

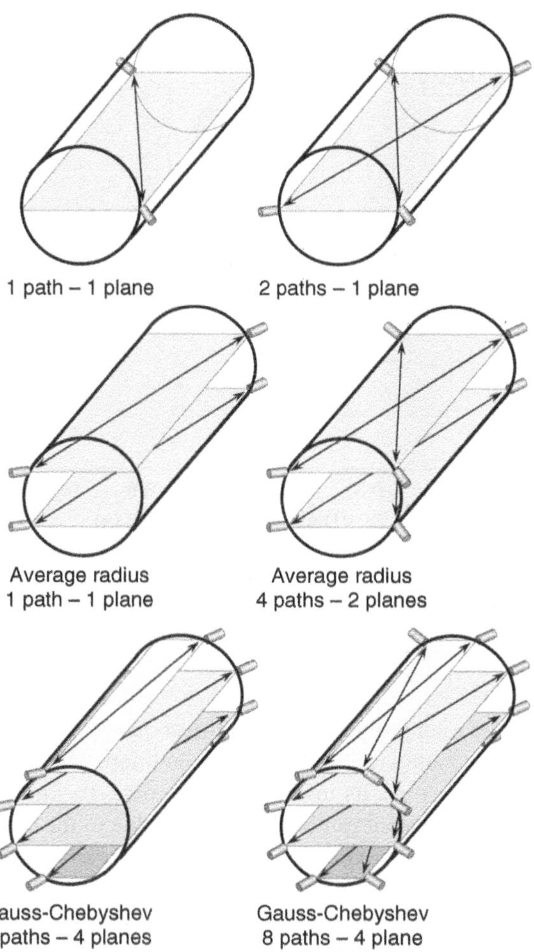

FIGURE 3.9.   EXAMPLES OF POSSIBLE CONFIGURATIONS IN THE PLACEMENT OF TRANSDUCERS

TABLE 3.1.   WEIGHTING COEFFICIENT FOR DIFFERENT TRANSDUCER
CONFIGURATIONS IN ULTRASONIC TRANSIT TIME FLOW METERS

Number of planes	Distance from the axis	Weighting coefficient
2	$+0.52\,R$	0.5
	$-0.52\,R$	0.5
	$+0.77\,R$	0.278
3	0	0.444
	$-0.77\,R$	0.278
	$+0.86\,R$	0.174
4	$+0.34\,R$	0.326
	$-0.34\,R$	0.326
	$-0.86\,R$	0.174

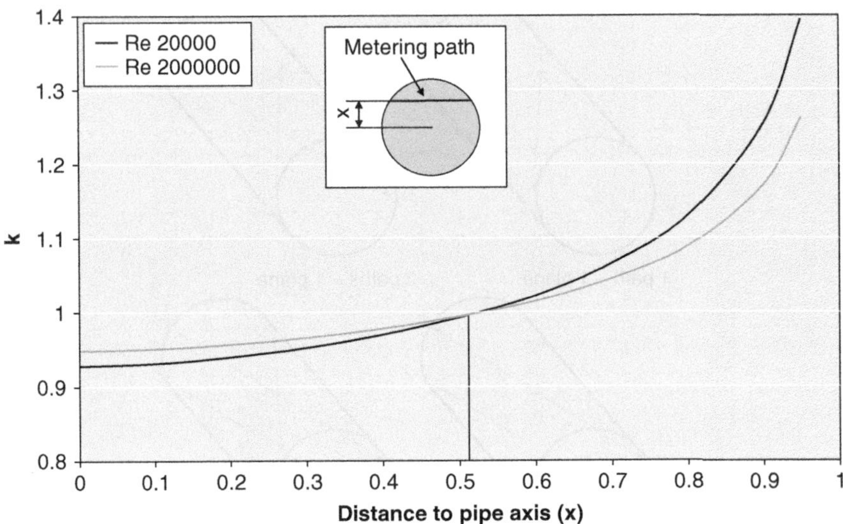

FIGURE 3.10.   RELATIONSHIP BETWEEN THE VELOCITY AT DIFFERENT PLANES AND THE AVERAGE VELOCITY IN THE PIPE FOR TWO REYNOLDS NUMBERS

On the other hand, clamp-on transducers reach accuracies ranging from 0.5% to 2% of rate. However, these figures refer to laboratory conditions where the internal diameter and the thickness of the pipe, its material, etc. are perfectly known. In practice, the uncertainty in the measurement can be much higher due to the lack of precise knowledge of the parameters needed to obtain the flowrate, especially the internal diameter.

Transit time meters are appropriate for pipes with diameters ranging from 10 to 2000 mm, or even larger. The velocity range for the device to provide adequate measurements is from 0.5 to 10 m/s, with the error increasing as velocity decreases.

Doppler effect ultrasonic flow meters

Doppler effect meters are rarely very accurate. The reason is that the meter does not really determine the velocity of the fluid, but rather the velocity of the suspended particles or bubbles. Sometimes these two velocities are quite different and can create important errors in the determination of the flowrate. In the best cases, the error of a Doppler meter will be around 2%.

3.1.6.   Metrological characteristics

Metrology of transit time meters: Velocity profile

As previously mentioned, ultrasonic meters measure velocity in a single or several planes and consequently are affected by distortions in the velocity profile. Depending on the number of transducers, the influence of the profile distortions will be more or less severe.

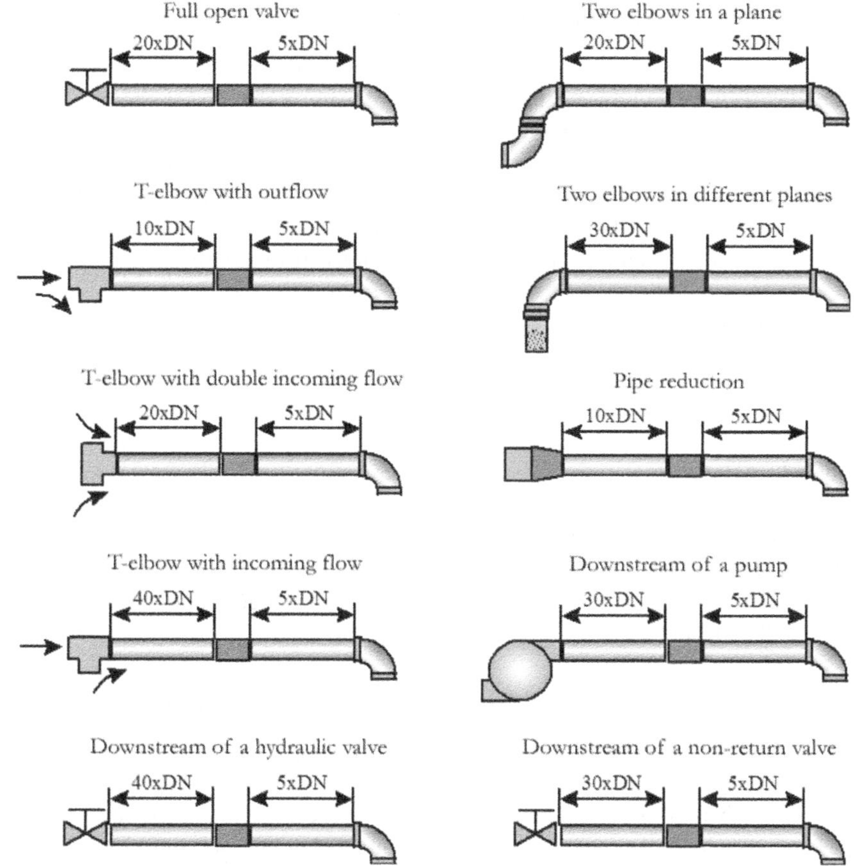

FIGURE 3.11. REQUIREMENTS OF THE INSTALLATION FOR ULTRASONIC METERS

In cases where only a pair of transducers are used (single path), it is important to guarantee that the quality of the velocity profile is as high as possible. In these cases, the recommended length of straight pipe upstream from the meter is the one shown in Figure 3.11. However, some authors recommend to increase these length to reduce uncertainty. Baker (2000) after reviewing the works of several authors makes the following recommendations to reduce uncertainties below 2%:

- After a reduction: 15 diameters.
- After an elbow or T: 20 diameters.
- After two or more elbows out of plane: 40 diameters.

If error was to be maintained below 1% in a flow meter with two paths the requirements would be as follows:

- After a reduction: 10 diameters.
- After an elbow or T: 10 diameters.
- After two or more elbows out of plane: 20 diameters.

Metrology of transit time meters: Suspended solids and air

The presence of air or suspended particles may block the transmission of the sound beams between the transducers, and consequently the use of this type of ultrasonic meters in dirty waters or waters with a high content of dissolved air is not recommended.

For this reason, the installation of sensors in a horizontal plane may avoid losses of signal due to the air present in the pipe, which usually accumulates in the high part of the pipe. However, this measurement will not be free from the errors that result from accounting the air as water and the distortions created in the velocity profile.

Metrology of transit time meters: Sedimentations

In the lower part of the pipe there are usually all types of deposits. These deposits are the cause of two problems. The first one and most important is that these deposits notably attenuate the signal. The second one is that with time, they may considerably reduce the cross section of the pipe in such a way that the flowrate is incorrectly estimated from the average velocity. While the first problem can be solved by placing the transducers on a horizontal plane, the second one is difficult to solve. The best solution in these cases is prevention, by forcing flow velocities to reach values of 2–6 m/s (since meters can measure up to 10 m/s) in the length of pipe where the flow meter will be installed.

Metrology of transit time meters: Additional considerations

When the transducers are externally clamped on the pipe, it is important to take into account the loss of signal that can derive from pipe materials (especially with porous materials such as asbestos cement, cast iron, concrete, etc.). Quite often, in addition to the heterogeneity of the material, the dirt accumulated on top of it must be considered in a possible loss of signal, since it prevents a correct acoustic coupling between the transducers and the pipe material. Consequently, it is important to clean the area where the transducer is going to be placed to ensure that the contact with the pipe is satisfactory.

In painted pipes, air and dirt build up between the paint layer and the pipe itself, attenuating and distorting the signal. As a consequence it is recommended that in this sort of pipes the paint is removed before placing the transducers (especially if accurate measurements are needed).

The measurement method must be chosen according to the diameter of the pipe. For instance, for small diameters the reflection method shown in Figure 3.12 should be used, in which transducers are placed on the same generatrix. With larger diameters, the sound waves should travel directly from one transducer to the other directly, without the reflection on the pipe walls.

The setup of the transducers in direct mode is, in the field, quite more complicated than in reflected mode. Both methods require to take into account the distance between the sensors (something easy to do with the supplied equipment). However, the direct method additionally requires to place them in an opposite generatrix (Figure 3.13).

In both cases, the distance on the longitudinal axis is calculated electronically taking into account variables such as the inner diameter, the thickness of the pipe, the speed of sound in the liquid, the incidence angle, etc.

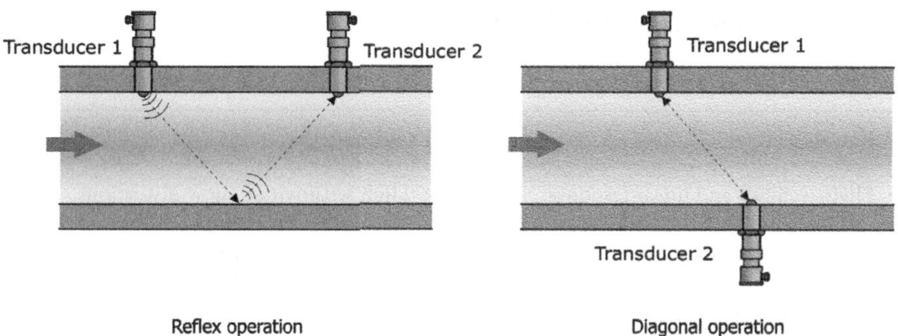

FIGURE 3.12. MEASUREMENT METHODS FOR TRANSIT TIME METERS DEPENDING ON THE PIPE DIAMETER

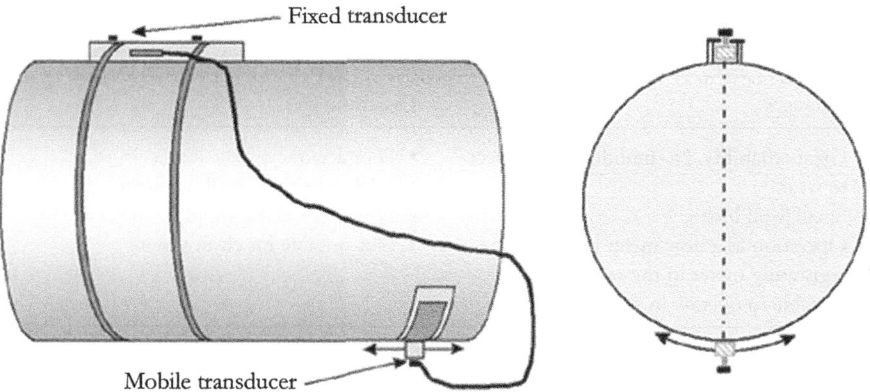

FIGURE 3.13. SETUP OF AN ULTRASONIC FLOW METER WITH CLAMP ON TRANSDUCERS

Clamp-on transducers allow the meter to be portable, but also allow to double check the quality of the velocity profile by carrying out several measurements in different planes of the pipe. If all measurements are the same for several rotated planes, the velocity profile will quite likely be symmetrical. However, if the measurements present discrepancies, the results should be handled cautiously.

Metrology of Doppler effect meters

Doppler effect flow meters are instruments of low accuracy that should only be used to get a rough estimation of the flowrate. Additionally, they are subject to greater inaccuracies due to sedimentations, or any distortion of the velocity profile. Dirt in the wall pipes will also increase the error in a meter which by design relies on the solids suspended in water.

Last but not least, meters intended to measure flowrates in open channel flow are prone to additional errors related to the measurement of the level of water or the section of the channel. These meters, when installed permanently, may also be affected by sedimentation and stop functioning without adequate maintenance.

3.1.7. Advantages and disadvantages

Transit time meters

Advantages	Disadvantages
• High accuracy and linear response. • Great reliability. No mobile parts subject to wear. • Small head losses. • Operation as a flow meter and a registering meter in the same device. • Moderate cost. Price is not linear with diameter increase.	• High sensitivity to flow distortions. • Need for electric supply. • Not suitable for dirty waters. • Not suitable for billing (at least for the time being). This may change in the future with the publication of newer standards.

Doppler effect meters

Advantages	Disadvantages
• Great reliability. No mobile parts subject to wear. • Small head losses. • Operation as a flow meter and a registering meter in the same device. • Suitable to operate in slurry waters.	• Poor accuracy. • High sensitivity to flow distortions. • Need for electric supply. • Not suitable for clean waters.

3.2. ELECTROMAGNETIC FLOW METERS

3.2.1. Operating principles

The operation of electromagnetic meters is based in Faraday's induction law. This law states that a voltage (E) appears between the ends of any conductor passing through a magnetic field. The induced voltage is directly proportional to flow velocity (V), the length of the conductor (L) and the intensity of the magnetic field (B) that surrounds it.

$$E = K \cdot B \cdot L \cdot V \tag{3.14}$$

In the case of an electromagnetic flow meter, the role of the conductor is played by the water or the fluid placed between the two electrodes. The length of the conductor would be approximately the distance between the electrodes, which in most cases is the diameter. The velocity would be a weighted average velocity of the fluid in the electrode section. The intensity of the magnetic field depends on the magnetic coils that the manufacturer places in the housing. The induced voltage can be measured by the electrodes that, in most cases, are in direct contact with the conductive medium (water) (Figure 3.14).

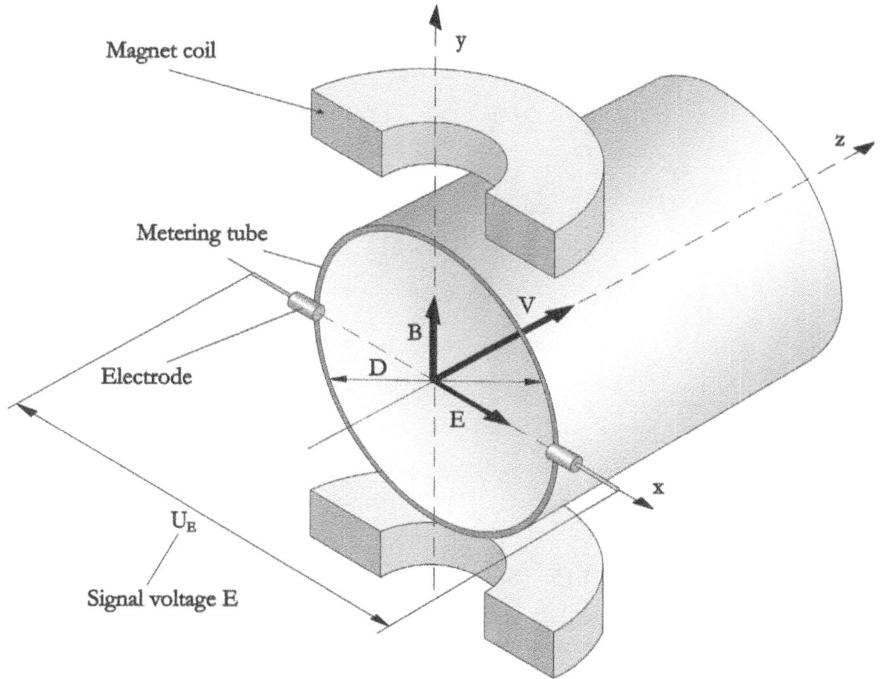

FIGURE 3.14. SCHEMATIC OF AN ELECTROMAGNETIC FLOW METER

Taking into account that the volumetric flowrate in a circular pipe may be obtained by the well-known expression:

$$Q = \frac{\pi \cdot D^2}{4} \cdot V \tag{3.15}$$

It may be combined with Faraday's Law to find out that the flowrate is proportional to the voltage (E) measured in the two electrodes, and inversely proportional to the intensity of the generated magnetic field. The proportionality constant C is particular to every meter:

$$Q = \frac{\pi \cdot D^2}{4 \cdot K \cdot L}\left(\frac{E}{B}\right) = C\left(\frac{E}{B}\right) \tag{3.16}$$

From the previous expressions it can be deduced that the induced voltage is not only a function of fluid velocity, but also depends on the intensity of the magnetic field B, which is also subject to variations of the voltage feeding the coils and the changes in conductivity of the fluid. This problem is solved by the manufacturers by creating another reference signal, which derives from the same magnetic field. By comparing both signals, the variations due to changes in the magnetic field or the fluid conductivity are discarded.

To understand electromagnetic flow meter working principle it is important to realize that in this instruments there is not a "single" conductor to which Faraday's law applies. The conductor in this case is water, and it circulates at a different velocity

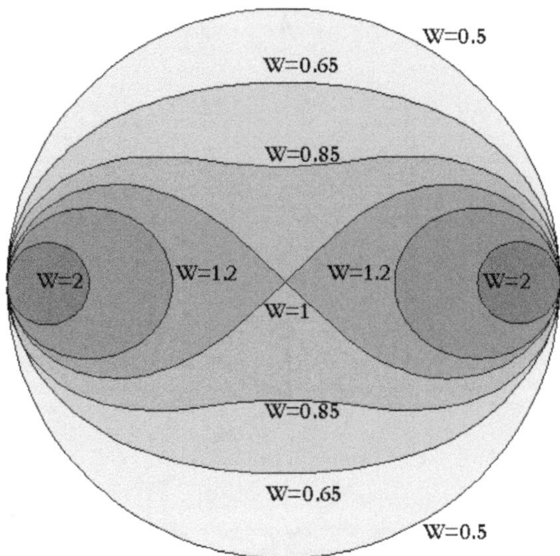

FIGURE 3.15.   SHERCLIFF WEIGHTING FUNCTION

depending on the section area considered. Additionally, not all areas in the section contribute in the same way to generate voltage between the electrodes. As a consequence, the value of the velocity to be used in the expressions previously mentioned has to be a weighted average. Shercliff (1962) developed a theoretical model by supposing uniform magnetic field to estimate the contribution of each area to the voltage, creating a weighting function (Figure 3.15). This function also allows to assess how certain flow distortions will affect accuracy.

The new electromagnetic meter designs try to incorporate coils and electrodes that intend to reduce the differences in the weights used in the function, making it more uniform and closer to one. This reduces the sensitivity of the meter to flow distortions. Details on this topic can be found in works by Baker (1982, 1983), Hemp and Sanderson (1981) and Al-Khazraji and Hemp (1980).

3.2.2.   Constructive characteristics

An electromagnetic flow meter is constituted by two components. The primary element is the one the water uses to flow through and where the sensors (electrodes) are installed. It provides as an output a weak electric signal (alternate current, AC) which is created at the electrodes in the pipe. The secondary element receives this signal, processes it and transforms it into a standard signal which can be transmitted to other devices. The latest models additionally register flow in both directions and record the extreme values, totalize flow, etc.

The primary element of an electromagnetic meter has the following components (Figure 3.16):

- One or several coils and a ferromagnetic nucleus, arranged in such a way that the generated magnetic field is perpendicular to the direction of the flow.

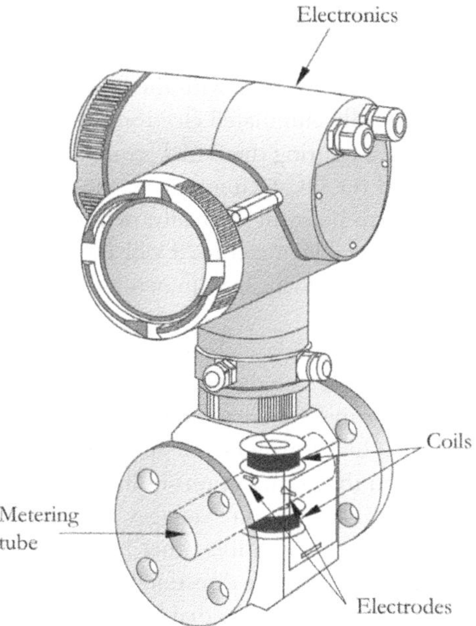

FIGURE 3.16.   **DETAIL OF THE PRIMARY ELEMENT OF AN ELECTROMAGNETIC FLOW METER**

- A metering tube of non-ferromagnetic material. The fluid flows through this tube. Should this tube be metallic, it has to be coated with non-conductive material to avoid the electrodes being in short circuit or otherwise provide some way to prevent a direct electric contact of the two electrodes.
- The electrodes which measure the induced voltage in the fluid section. The voltage detected by these elements is sent to the secondary module for its amplification and treatment.

Electromagnetic flow meters can be classified according to the type of current (AC or direct current, DC) feeding the coils. DC meters are more modern and appropriate for most uses than AC ones. However, in a reduced number of applications an AC meter may be more suitable than a DC one. Such is the case of measures with fluids with a large quantity of entrapped air, non-homogeneous suspended particles, or in cases where positive displacement pumps with frequencies lower than 15 Hz are used. In these situations, DC meters may provide an output signal with excessive noise.

If the coils are fed with AC, the induced voltage becomes sinusoidal and proportional to the voltage and frequency of the feeding current (besides being proportional to the fluid's velocity). Meters with this construction usually use a reference signal derived from the same magnetic field to cancel the effects of small variations in the voltage and frequency of the feeding current.

Faraday's law also applies to static conductors subject to variable magnetic fields. This is the case of the cables connecting the electrodes and the secondary element. They act as stationary conductors subject to a variable magnetic field, either sinusoidal (AC meters) or with a pulsating DC signal (DC meters). As a consequence the secondary element will receive two signals. One from the primary element which is proportional to the flowrate, and another one from the static conductors considered as "noise".

This noise may or may not be in phase with the flowrate signal. The part of noise out of phase (up to 90°) is easily eliminated electronically. The component in phase can be isolated at zero flowrate during the initial setup.

One of the advantages of the DC technology vs. the AC one is that these meters do not need to adjust the zero position manually, neither during the initial installation nor during the regular use. The meter itself will automatically perform this task. The need of a regular zero adjustment is a known and substantial inconvenient of AC electromagnetic meters.

The signal received in the secondary element is treated electronically (both the electrodes' signal and the reference signal). The output signal is standard (for instance, 4–20 mA, and 0–10,000 Hz in frequency) and proportional to the flowrate. Additionally, this element adjusts to compensate the variations in tension and frequency of the coils' feeding current and the perturbations in the intensity of the magnetic field in the primary element, eliminates parasite voltages, etc. As the technology becomes more sophisticated, this element also fulfils other functions, such as cancelling the signal at very low flowrates, detecting the direction for the flow, integrating the circulating flows to obtain values for the volumes that have circulated in each direction, etc.

Finally, and regarding the metering tube, the manufacturers offer the possibility of different coatings for it. The ISO 6817:1992 contains an informative annex listing several coating materials for the primary element, and their qualities.

Coating of the measuring tube

- *Hard rubber or ebonite*: High temperatures between 0°C and 100°C. Good resistance against abrasion originated by small size particles. Good chemical resistance against lixiviation, acids and bases.
- *Natural rubber*: Temperatures from −20°C to 70°C. Good abrasion and chemical resistance.
- *Polyurethane*: Temperatures from −50°C to 50°C. Good impact and wear resistance.
- *Polytetrafluoroethylene (PTFE)*: Normally used as an extruded lining which is not adhered to the tube. Temperatures ranging from −0°C to 200°C. Excellent wear resistance to small particles, and chemically inert. Resistance may be reduced at pressures lower than the atmospheric.
- *Polyamides*: Temperatures lower than 65°C. Good resistance to wear.
- *Glass fibre-reinforced plastic (GRP)*: Temperatures ranging from 20°C to 55°C. It can be used both for coating and construction of the metering tube. Especially appropriate for large diameter primary elements.
- *Ceramics*: Temperatures from −60°C to 250°C. This material does not require an inner coating. It is very resistant to pressure and temperature variations and to abrasion. Some ceramic materials are specially stable against acids and alkaline solutions.
- *Teflon/TFA*: Corrosion and temperature resistant. Moderate resistance to abrasion.

The electrodes' material is also important depending on the characteristics of the fluid. The most usual materials are stainless steel (for non-corrosive liquids) or platinum, platinum/iridium, titanium and some alloys for corrosive liquids.

The meter electrodes require to be cleaned with a periodicity which depends on the characteristics of water. Several cleaning methods can be used:

- Extracting the electrodes while the meter is in service.
- Burning off any deposits in the electrodes with an electric current of sufficient intensity.
- Using ultrasonic pulses. In this technique, the electrode plays the transducer role, vibrating and eliminating the deposits.

3.2.3.   Metrological characteristics and dimensions

Electromagnetic flow meters are one of the most precise devices to meter flows that can be found in the market nowadays. Their accuracy usually ranges between 0.5% and 0.1% of rate, values which are significantly lower than those found for water meters.

These values are conditioned by the flow velocity. As a matter of fact, their accuracy is a function of the difference in potential between the electrodes, and the smaller the velocity, the smaller this voltage is, increasing the error to determine it. As a consequence, the error curve shows how a reduction in the velocity implies an increase in the error. The recommended lower limit for most models is 0.5 m/s. When the velocity presents lower values, the error is increased proportionally to the decrease in velocity as shown in Figure 3.17.

Like any other instrument, an electromagnetic meter must be dimensioned as a function of its flow capacity and not of its diameter. Usually, if the installed meter matches the pipe's diameter velocities are too low and the error is too large. Table 3.2 shows the recommended flowrates for each diameter and error.

The sizes of commercial electromagnetic meters range from 2 to 3000 mm, although they are usually found in sizes from 60 to 500 mm.

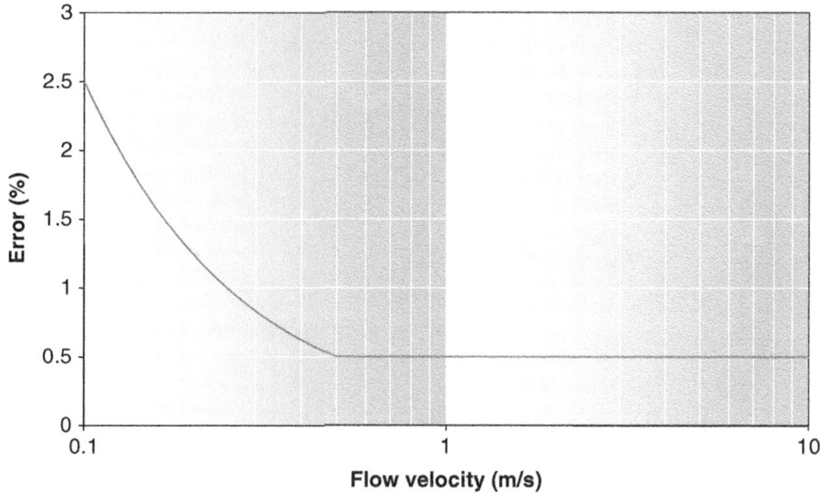

FIGURE 3.17.   ERROR CURVE OF AN ELECTROMAGNETIC METER AS A FUNCTION OF THE FLUID'S VELOCITY

TABLE 3.2. RECOMMENDED FLOWRATE AS A FUNCTION OF DIAMETER AND ACCURACY

Diameter (mm)	Maximum flowrate (m^3/h)	Minimum flowrate (m^3/h)	
		Error <0.5%	Error <2.5%
15	6.4	0.32	0.06
25	18	0.88	0.18
40	45	2.26	0.45
50	71	3.53	0.71
65	119	5.97	1.19
80	181	9.05	1.81
100	283	14.1	2.83
125	442	22.1	4.42
150	636	31.8	6.36
200	1131	56.5	11.31
250	1767	88.4	17.67
300	2545	127	25.45
400	4524	226	45.24
500	7069	353	70.69
600	10,179	509	101.79
800	18,096	905	180.96
1000	28,274	1414	282.74

FIGURE 3.18. LARGE ELECTROMAGNETIC METER (1800 MM)

3.2.4. Metrology of electromagnetic meters

Metrology of electromagnetic meters: Conductivity of the fluid

To ensure the correct operation of an electromagnetic meter, the fluid's conductivity must be greater than $2\,\mu S/cm$. However, there are certain electromagnetic meters in which the electrodes are not in direct contact with the fluid and the induced voltage signal is picked up by capacitive coupling. This allows for a larger sensitivity of the electrodes, and as a consequence a lower conductivity of up to $0.05\,\mu S/cm$ is tolerated. Complying with these values is not a real problem in metering drinking or irrigation water, with an approximate conductivity of $100\,\mu S/cm$.

Meters with AC technology may present offset errors when the conductivity changes suddenly. However, DC models should present no problems as long as the conductivity of the fluid is uniform throughout the section and above the threshold value of the device. The reason for this is that if the conductivity of the fluid is uniform for the whole section of the metering tube, the distribution of the electric field is independent of the conductivity and as a consequence so will be the signal.

Should the conductivity not be uniform, the following considerations should be taken into account. An electromagnetic flow meter measures the velocity of the components with a higher conductivity flowing through it, and calculates the flowrate taking into account the cross section of the meter. For instance, if water is dragging some suspended particles such as air or sand, the relative conductivity of these elements will decide which velocity is measured. Air and sand with almost no conductivity will not present a problem, and the velocity of water will be used to calculate the flowrate. However, if the concentration of these other elements is high, the meter will be inferring that everything flowing through the meter is water, and consequently the error will be considerable.

Furthermore, a fluid of a non-homogeneous conductivity will alter the weighting function, modifying the contribution on the induced voltage of every point in the cross section.

One factor which influences conductivity of water is the dilution of chemical products. If these products are injected upstream from the meter, the conductivity values may be affected considerably, creating an important amount of noise in the output signal, caused by the heterogeneity of the conductivity of the fluid in the electrode section. Therefore it is recommended that, if there is the need to add chemicals to the water, it is done at a sufficient distance to guarantee that the product is totally diluted and does not affect the measure.

Metrology of electromagnetic meters: Velocity profile

Electromagnetic meters are sensitive to velocity profile distortions, although this sensitivity is lower than for other types of meters. In most applications, a distance of five diameters of straight pipe upstream of the meter and three diameters downstream is enough to guarantee correct measurements (studies show that a shorter downstream distance may produce errors up to 5%). If the meter is installed downstream from a pump, the distance must be increased to eight diameters from the distorting element.

Metrology of electromagnetic meters: Suspended solids

The use of electromagnetic meters with fluids in which there are abrasive solids in suspension may result in wear in particular areas (especially with certain velocity profiles, such as the one created if an elbow is placed upstream from the meter). This effect can be minimized by placing a diaphragm at the upstream joint, but always taking into account how this device will affect accuracy.

In such cases, the selection of the inner coating of the metering tube is vital to ensure a long lifespan for the meter. The most important factors to consider while choosing the coating are the chemical composition of the fluid, the abrasive characteristics and the operating temperature and pressure.

The diameter of the electromagnetic meter must allow a velocity ranging from 1 to 5 m/s in normal operating conditions. Higher velocities may lead to an excessive wear of the coating. Lower velocities may favour solid depositions.

Metrology of electromagnetic meters: Limescale build-up

One of the most frequent problems found in electromagnetic meters is limescale build-up in the electrodes and the tube, which may even lead to reducing considerably the cross section of the meter. The problem with the electrodes is relatively important, especially as long as they maintain electrical contact with the fluid.

Since the flowrate is obtained by multiplying the original cross section by the velocity, a reduction in the section will increase the measured flowrate (the velocity measured will be increased, but only because the section has been reduced without the meter "knowing").

Avoiding limescale build-up is often almost impossible, even though manufacturers devote quite a lot of effort to this task. In any case, the best solution is always prevention. The meter should be set up in such a way that the velocity of the fluid through the tube is between 2 and 4 m/s, and as fast as possible within this range while avoiding excessive wear. In order to reach these velocities, the flow meter often needs to be of a lesser diameter with respect to the pipe, needing conical reductions (usually of 8° in angle). Higher velocities create more turbulence around the electrode and reduce limescale build-up in electrodes and tubes.

Metrology of electromagnetic meters: Entrapped air

The presence of air in the pipes may also be a cause for metering errors, especially when the air bubbles are not distributed uniformly and air pockets are created. The best solution to this problem is to create a very turbulent flow in the meter in order to uniformly distribute the existing air. However, this solution does not solve the problem that the air will be accounted for as water, thus slightly increasing the metered flowrate (with respect to the real amount of water flowing through the meter).

Baker and Deacon (1983) performed a series of experiments to determine the behaviour of an electromagnetic meter, with a vertical setup, when measuring a mixture of water and air. With a fraction of air of up to 8%, the metering error was less than -1%. However, the magnitude of the error increased with the amount of entrapped air.

3.2.5. Channel flow metering

Nowadays some electromagnetic meters in the market are able to measure flowrates in pipes which are not pressurized. The design of such meters takes into account two important aspects. First, the velocity profile for channel flow is completely different to the one found in pressurized flow. Additionally, it is necessary to measure the depth of water in order to assess the circulating flowrate. Consequently these meters often incorporate a level sensor.

The accuracy of these devices when measuring channel flow is much smaller than the one found for pressurized flow, and often is about 1% of the upper range value.

3.2.6. Grounding of the meter

The liquid to be measured and the housing of the primary element should have the same electrical potential and be grounded. As a matter of fact, the process of grounding these types of meters represents one of the key steps in the setup of the meter in order to obtain a noise free signal. The grounding of a meter depends on the electrical properties of pipe. In conducting pipes, the pipe and meter flanges should be connected electrically. If the pipe cannot be considered a proper "ground", the flanges should also be grounded with a cable of enough section and length as short as possible (Figure 3.19).

When installed in a non-conducting pipe, the meter should be equipped with grounding rings. These metallic rings have the same inner diameter as the nominal diameter of the meter (to avoid turbulences) and must be placed between the flanges of meter and pipe. They should be connected both to the meter flanges and an appropriate ground. These rings may be quite expensive for large diameters.

Some meters included a third electrode, frequently named earthing electrode, which provide a reference potential for calculating the flow signal. This extra electrode makes unnecessary the use of grounding rings in order to bring the same electric potential to the electronics of the meter and the fluid.

3.2.7. Installation conditions

In normal operating conditions, an electromagnetic meter should always have the metering tube full of water. Otherwise the output signal will be erratic and unpredictable. Most manufacturers offer an option in which the secondary element sends a warning signal or stops measuring when the metering tube is not running full. Other flow meters are equipped with electronics that continuously monitor the fluid conductivity in the vicinity of the electrode, in order to block the output signal should this one reach a certain limit (for instance showing that the electrodes are in contact with air). However, this type of alerting system may not work properly for fluid conductivities below $20\,\mu S/cm$.

In order to guarantee that the metering tube is always full, it is recommended to install it in a vertical pipe with ascending flow. However, horizontal setups are very common, and in such cases it should be ensured that the electrodes are not placed in

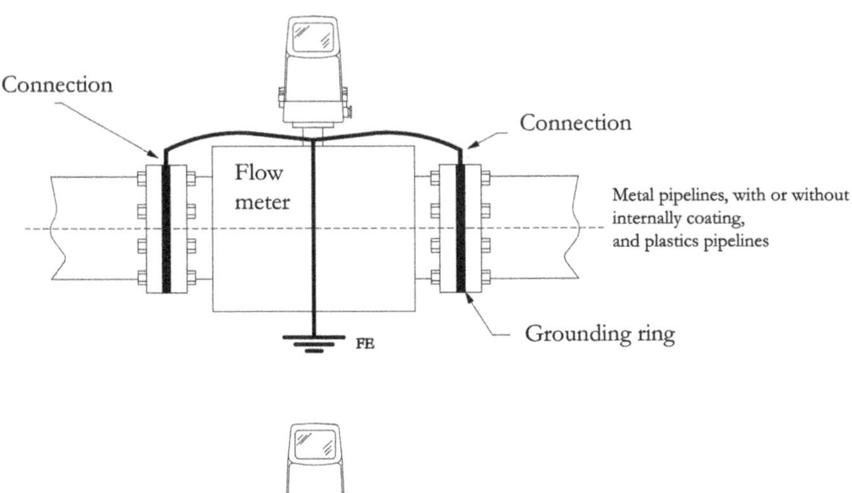

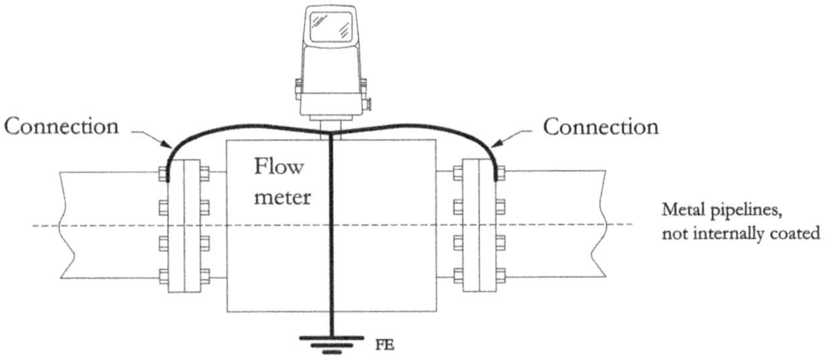

FIGURE 3.19.   CONNECTIONS TO ENSURE THE ELECTRICAL CONTINUITY IN FLANGES

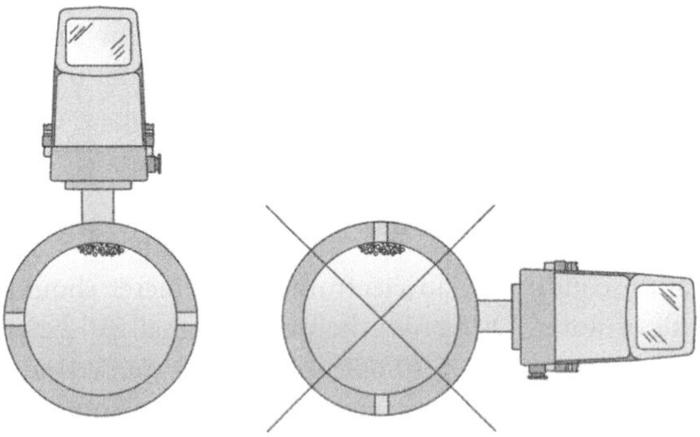

FIGURE 3.20.   ELECTRODE ORIENTATION

a vertical plane (Figure 3.20) to avoid potential electrical isolation of the electrodes by air pockets and also the accumulation of solids around them.

Figures 3.22 and 3.23 show correct configurations for the installation of electromagnetic meters. As a rule of thumb, the meters should be placed in ascending or

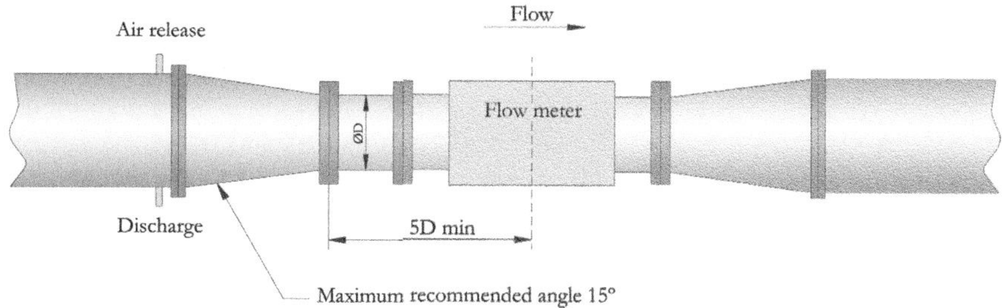

FIGURE 3.21. SETUP OF AN ELECTROMAGNETIC METER FOR A LARGE-DIAMETER PIPE

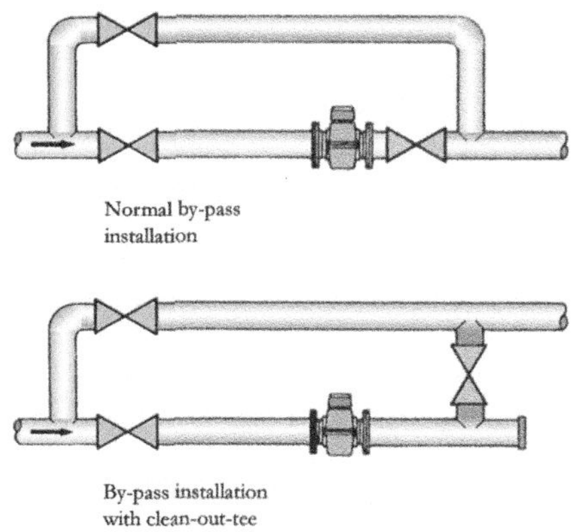

FIGURE 3.22. BY-PASS INSTALLATION OF AN ELECTROMAGNETIC FLOW METER

vertical pipes which will always be full of liquid. Electromagnetic meters should never be installed at a pump intake or immediately downstream from a regulation valve.

3.2.8.   Advantages and disadvantages

Advantages	Disadvantages
• High accuracy and linear response. • Great reliability. No mobile parts subject to wear. • Small friction losses. Low operation cost. • Can register flowrate and volume in one instrument. • Can be used in loaded waters. • Low sensitivity to flow profile distortions • Moderate cost.	• Need for electric supply. Some devices available in the market are autonomous and can operate without replacing the batteries up to 2 years. • Need to be grounded. • Cannot legally be used for billing, although new standards will change this in the future. • Need protection against storms. Failures due to thunderbolt are frequent.

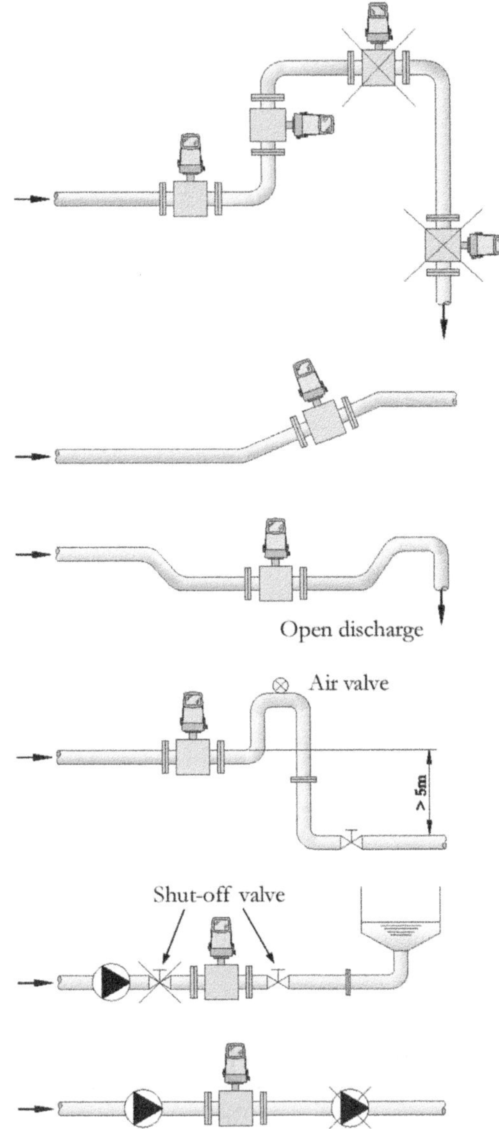

FIGURE 3.23.   INSTALLATION RECOMENDATIONS FOR ELECTROMAGNETIC FLOW METERS

3.3.   INSERTION FLOW METERS

3.3.1.   Operating principles

Although nowadays there are many types of insertion probes with different operating principles, the most common ones are electromagnetic probes, turbine probes and differential pressure probes (Pitot tubes).

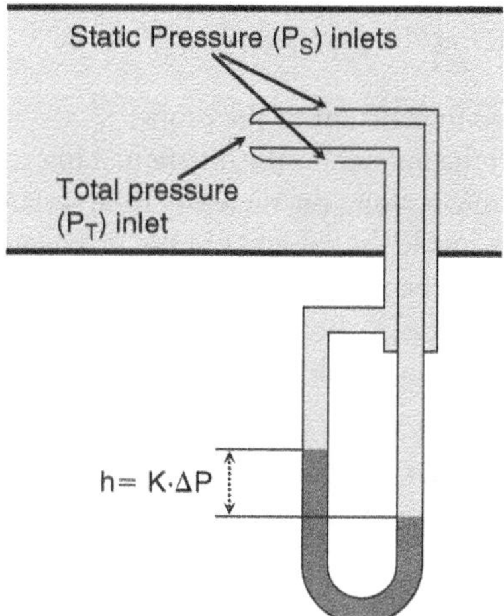

FIGURE 3.24.   PITOT TUBE

These probes are often used to other meters calibrate on-site or, due to their low cost, as a portable meter which can be used in different points of a distribution network. However, the most common uses of these type of devices may not always be the most appropriate ones.

With independence of the operating principle, all these probes aim to measure the local velocity of the fluid. The techniques used to this end are several: exchange of momentum for turbines, induced voltage picked up (electromagnetic) or the estimation of the kinetic term of the flow (Pitot tubes, averaging Pitot probes).

Differential pressure probes

Pitot tubes are able to determine the local velocity at the tube's intake. This is achieved by comparing the pressure at this point (total pressure) with the static pressure which does not include the kinetic term. The difference is, as a matter of fact, the kinetic term in the front opening of the tube. For a non-compressible fluid (Figure 3.24):

$$\text{Total pressure } (P_{\mathrm{T}}): \quad \frac{P}{\gamma} + z + \frac{V^2}{2g} \tag{3.17}$$

$$\text{Static pressure } (P_{\mathrm{S}}): \quad \frac{P}{\gamma} + z \tag{3.18}$$

$$\text{Pressure difference:} \quad \Delta P = P_{\mathrm{T}} - P_{\mathrm{S}} \tag{3.19}$$

$$\text{Theoretical velocity at the tube's front intake:} \quad V = \sqrt{2 \cdot g \cdot \Delta P} \tag{3.20}$$

Actual average velocity at the pipe section: $V = K \cdot \sqrt{2 \cdot g \cdot \Delta P}$ (3.21)

with K being the calibration constant of the probe.

There are variations in the market for the differential insertion probes called "averaging Pitot tubes" which present along the metering tube several orifices. The purpose is to connect all this points inside the tube to obtain an average value of the total pressure of the fluid impacting the sensor. By measuring this total pressure in different points, the influence of the velocity profile is reduced. However, these instruments are likely to produce inaccurate results unless they are handled by expert staff.

Electromagnetic insertion probes

These probes measure the velocity of the fluid following the same physical principle as electromagnetic meters, Faraday's law. The instrument generates a magnetic field by means of coils placed inside the tip of the probe. The fluid generates a difference of potential between the electrodes which is proportional to its velocity in this area near the sensor. Basically, like the other insertion probes, the flowrate is calculated from this value taking into account the section of the pipe and some other parameters.

The requirements with regards to the characteristics of the fluid are the same as those for full bore electromagnetic meters. The fluid's conductivity must be higher than $50\,\mu S/cm$.

Similarly to the differential pressure probes, it is possible to find special constructions with several sensors placed along the probe.

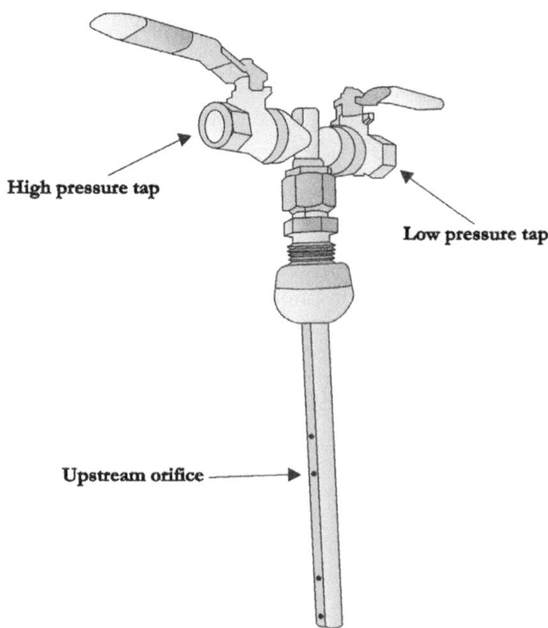

FIGURE 3.25.   AVERAGING PITOT TUBE

Turbine insertion probes

These probes determine the local velocity of the fluid from the rotational speed of a turbine. The rotation axis is frequently perpendicular to the pipe's direction, working like a Pelton wheel. However, some designs have a rotation axis perpendicular to the pipe's axis. The rotation of the turbine is measured by magnetic sensors, recording each turn of the wheel. Figure 3.27 shows a turbine insertion probe.

3.3.2. Constructive characteristics

In principle, and despite the particular characteristics inherent to each technology, the basic components of an insertion probe are:

- Velocity sensor, based on the technology of the probe.
- Isolation valve, used to install the device in a pressurized pipe.

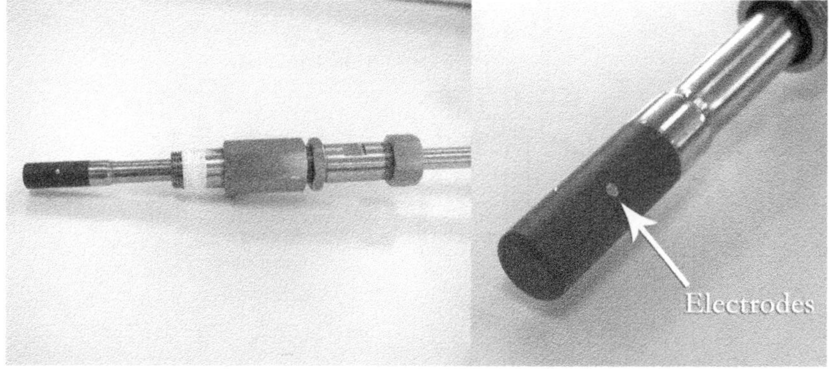

FIGURE 3.26.   ELECTROMAGNETIC INSERTION PROBE

FIGURE 3.27.   TURBINE INSERTION PROBE

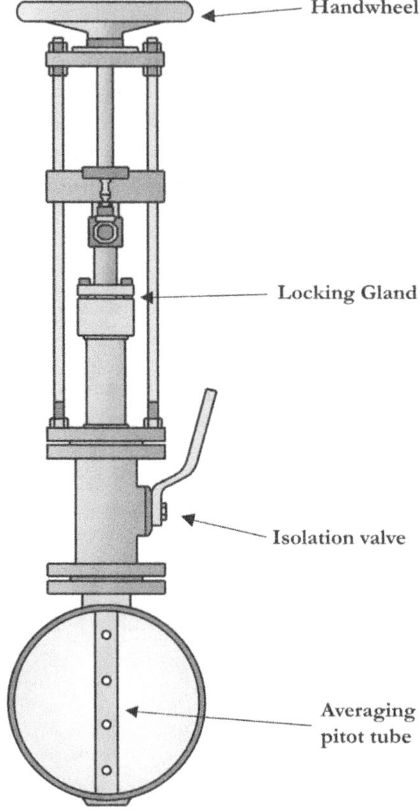

FIGURE 3.28.   ELEMENTS OF AN INSERTION PROBE

- Adjusting device to control the degree of penetration of the probe in the pipe.
- Locking gland to avoid that the internal pressure of the pipe expels the probe.
- Secondary element, transforming the sensor signal into a flowrate.

3.3.3.   Metrological characteristics and dimensions

The accuracy of insertion probes is not too high, although many manufacturers claim excellent theoretical values. The lack of reliability of these meters can be explained by the following facts:

- These instruments are very sensitive to velocity profile distortions. In practice, in the field, it is difficult to guarantee a fully developed velocity profile not leading to errors during flow measurements.
- Narrow measuring range, with a turndown ratio oscillating between 1:4 and 1:10. Additionally, the real accuracy of the probe will be affected by the capacity of the secondary element.
- Difficulty in installing correctly the device, especially with respect to the insertion depth and the orientation of the probe.

- Uncertainty of the actual internal cross section of the pipe. Limescale build-up or the deformation of the pipe may reduce the original internal cross section.

The velocity range for these devices goes from 1 to 5 m/s. This range is often limited by the capacity of the secondary element, for instance a differential pressure sensor, and by the mechanical resistance of the probe. It must be taken into account that in large diameters the torque created by the flow can be quite important. Additionally, if the frequency of the vortexes downstream of the probe happens to be the same as its natural frequency, the probe will most likely break.

As a consequence, the maximum velocity that a probe with a single support point can withstand depends exclusively on the inserted length. The larger the diameter, the smaller the velocity that the probe can withstand. For small insertion lengths the maximum velocity usually reaches 5 m/s. However, for insertion lengths over 1000 mm the maximum velocity is only 1 m/s.

Frequently the laboratory accuracy quoted by the manufacturers ranges between 5% and 1% of the measured velocity. The real accuracy strongly depends on the characteristics of the place where the instrument is mounted and the care taken during its installation. Measuring errors can be as high as 60% or more if the probe is not properly installed. However, for turbine and electromagnetic insertion probes, the precision values can be as low as 0.2% of the measure. This means that if a probe is maintained mounted in a given point and the flow conditions are the same, measured values will not differ in more then a 0.2%.

3.3.4.   Metrology of insertion flow meters

Metrology of insertion flow meters: Flowrate estimation

Insertion flow meters estimate the water flowrate by measuring the velocity in one point of the pipe. The location of the measuring points becomes extremely important, for the velocity of the water depends on the point in the pipe where it is measured (and as a consequence the relationship between this local velocity and the average velocity for the section is not constant if the location of the measuring point changes).

The techniques to determine the flowrate with insertion probes are basically three:

1. Velocity measured at the pipe's axis.
2. Velocity measured with an insertion depth of 1/8D or 7/8D.
3. Velocity measured at several points of a diameter.

Velocity measured at the pipe's axis   Consider the following equation which allows to obtain, for turbulent fully developed flow and a circular pipe, the velocity at each point of the section:

$$V_\mathrm{i} = \left(1 - \frac{r}{R}\right)^{1/n} \cdot V_\mathrm{max} \tag{3.22}$$

with n being a parameter dependant on the Reynolds number that can be calculated using the expression proposed by Nikuradse:

$$\frac{1}{n} = 0.2525 - 0.0229 \cdot \log(\text{Re}) \tag{3.23}$$

Taking into account that the average velocity in a pipe is obtained by integrating the velocity (Equation (3.22)) and dividing by the cross section of the pipe, the following expression relates the average velocity with the velocity at the axis:

$$V_{\text{avg}} = \frac{2 \cdot n^2 \cdot V_{\max}}{(1 + 2 \cdot n) \cdot (1 + n)} = F_{\text{p}} \cdot V_{\max} \tag{3.24}$$

This relationship is not constant and depends on the flowrate through the Von Karman coefficient, n. The coefficient F_{p} known as the profile factor, relating both velocities ranges from 0.79 for a Reynolds number of 10,000 and 0.875 for a Reynolds of 10^7 (Figure 3.29).

The fact that F_{p}, the proportionality constant between the average and maximum velocity is not stable and independent of the flowrate, presents some difficulties that can be solved by incorporating such coefficient to the formula used to calculate the flowrate.

In this case, the flowrate can be obtained from:

$$Q = F_{\text{p}} F_{\text{i}} F_{\text{o}} \cdot \text{Area} \cdot V_{\text{axis}} \tag{3.25}$$

where V_{axis} is the maximum velocity at the pipe axis, F_{p} is the profile factor, F_{i} is the insertion factor (taking into account the distortion in the velocity profile produced by the probe itself), F_{o} is the obstruction factor taking into account the proportion of the area blocked by the probe, which depends on the probe design and the diameter of the pipe.

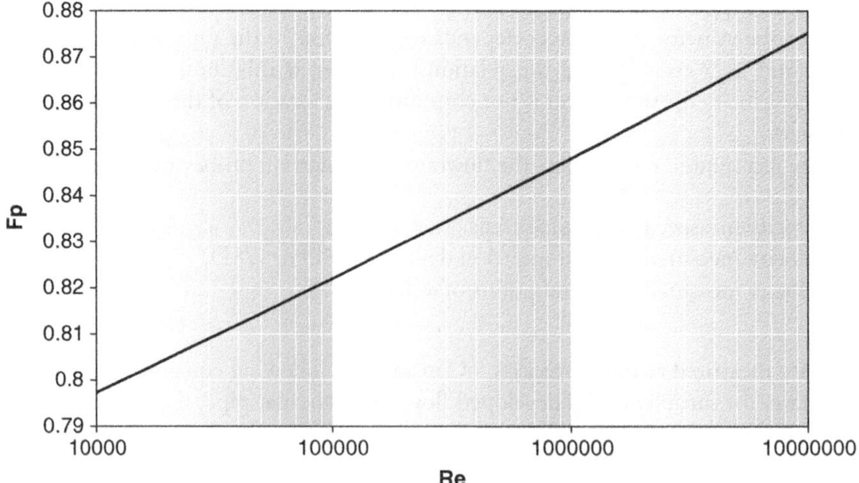

FIGURE 3.29.   RELATIONSHIP BETWEEN THE AVERAGE AND MAXIMUM VELOCITIES IN A CIRCULAR PIPE WITH FULLY DEVELOPED TURBULENT FLOW

Velocity measured with an insertion depth of 1/8D or 7/8D   In a fully developed turbulent flow, the velocity in a ring of radius 3/4R, is very close to the average velocity in the section, with independence of the value of the Reynolds number or the circulating flowrate (Figure 3.31).

The variation between the velocity in such ring $V_{3/4R}$, and the average velocity in the section V_{avg}, oscillates between 1.0035 and 1.0055 for Reynolds numbers between 10^4 and 10^7, which represents an error of only 0.2% (Figure 3.31).

Using this fact, the profile factor can be considered constant and equal to 1.005, allowing the calculation of the flowrate by means of the following equation:

$$Q = 1.005 \cdot F_i F_o \cdot \text{Area} \cdot V_{3/4R} \tag{3.26}$$

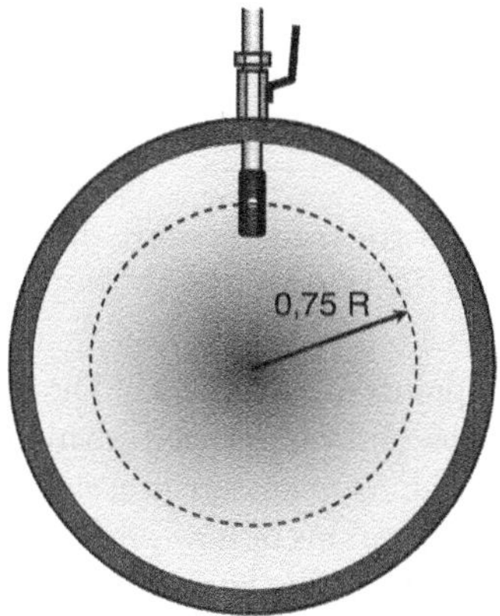

FIGURE 3.30.   VELOCITY MEASUREMENT AT 1/8D

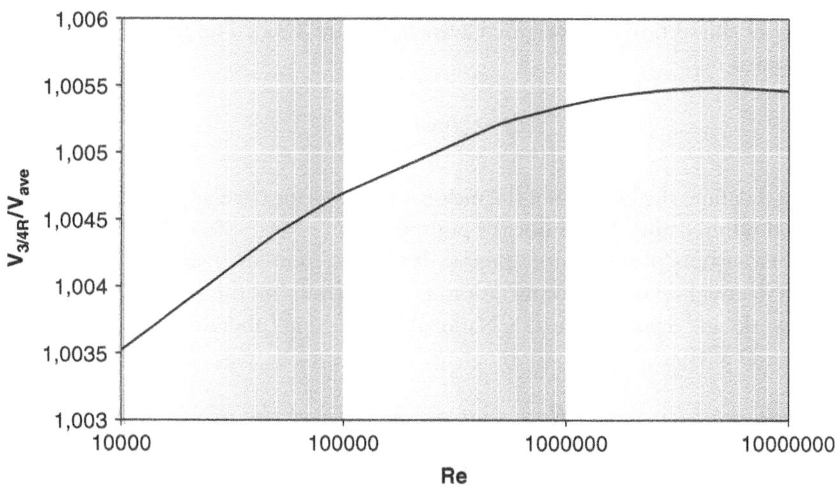

FIGURE 3.31.   ¾ R VELOCITY WITH RESPECT TO THE AVERAGE VELOCITY AS A FUNCTION OF REYNOLDS NUMBER

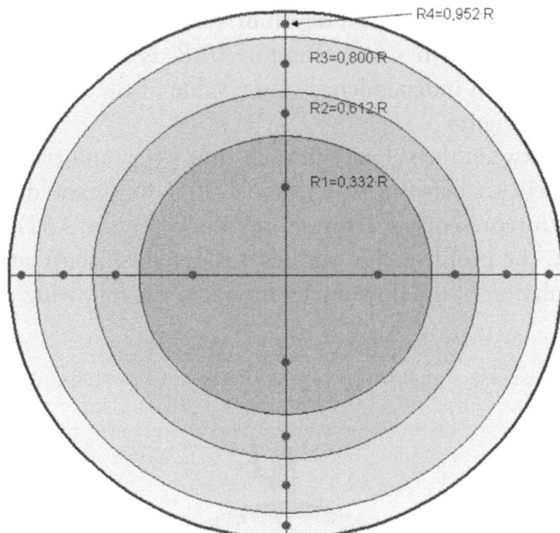

FIGURE 3.32.   EXAMPLE OF LOCATION OF EIGHT MEASURING POINTS WITH THE
LOG-TCHEBYCHEFF METHOD

Velocity measured in several points of a diameter   The previous two methods require a fully developed turbulent flow and without perturbations. Quite often these are not the conditions found in the field. As a consequence, alternative methods must be used to avoid the influence of the alterations of flow profile distortions.

Instead of measuring in a unique point, by measuring the velocity at several points of the section, the uncertainty is reduced. However, these points must also be carefully chosen. There are several methods proposed in the technical literature. Here, two of them are described: equal area centroids configuration or log-Tchebycheff configuration. By measuring the velocity in these points, it is possible to calculate the average velocity in the section as a weighted average (Figure 3.32). The flowrate would then be obtained by:

$$Q = \text{Area} \cdot \frac{1}{n} \sum_{i=1}^{n} V_i \tag{3.27}$$

Table 3.3 defines the location of the measuring points for circular pipes, depending on the method used and the number of points.

Other authors propose more precise, but more complex, methods of integration such as the method of cubic integration. Further details on this method can be found in the works and reports of the UK National Engineering Laboratory (report 290/2001, project FDWM04).

Metrology of insertion flow meters: Section of the pipe

Limescale build-up may reduce the real section of the pipe and affect the calculation of the flowrate considerably. Additionally, the variations in the rugosity of the pipe affect the velocity profile, and may change the relationship between the measured velocity and the average velocity in the section.

TABLE 3.3. LOCATION OF THE MEASURING POINTS IN DIAMETER PATHS

Points of measurement per path	Distance to the axis from the measurement point (r/R)					
Equal area centroids method						
6	±0.408	±0.706	±0.914			
8	±0.354	±0.612	±0.790	±0.936		
10	±0.316	±0.548	±0.708	±0.836	±0.948	
12	±0.288	±0.500	±0.646	±0.764	±0.866	±0.958
Log-Tchebycheff method						
6	±0.376	±0.724	±0.936			
8	±0.332	±0.612	±0.800	±0.952		
10	±0.286	±0.590	±0.690	±0.848	±0.962	

Metrology of insertion flow meters: Depth of insertion of the sensor

When the flowrate is calculated by measuring at a single point, the depth of insertion is a critical aspect in getting a reliable measure. As several works have shown (Cascetta et al., 2003) the values of flowrate obtained by measuring the maximum velocity at the axis are less sensitive to the correct insertion of the probe. However, when the velocity is obtained at the critical crown (3/4R) the errors can become significant.

Metrology of insertion flow meters: Determination of the local velocity

Insertion probes, like any other sensor, are not completely reliable when providing a measure of the real velocity. Occasionally, they also need to be calibrated, although the absence of moving parts, electromagnetic probes, increases the stability of the calibration. In any case, this does not guarantee a completely accurate measurement of the velocity. Furthermore, since this type of instruments are intended for field use, it is important to carry out a well-established maintenance program.

Metrology of insertion flow meters: Probe alignment and orientation

The net torque created by the impact of the water in a turbine insertion meter depends on the alignment of the velocity vectors of the water and the rotation axis of the turbine. The same happens with electromagnetic meters, where the voltage induced depends on the angle that forms the local velocity and the electromagnetic field. Consequently, alignment does affect the accuracy of the measure and should be considered for the installation of the probe (Figure 3.33).

The orientation of the probe also affects the metrology of the flow meter. All insertion probes require to be inserted perpendicularly to the pipe. Even slight deviations in any direction may affect the accuracy of the measurement. In any case, users should always check manufacturer recommendations.

Metrology of insertion flow meters: Velocity profile

All insertion probes deduce the value of the average velocity in the section of the pipe by measuring the velocity at one or more points in the pipe. Unlike an electromagnetic

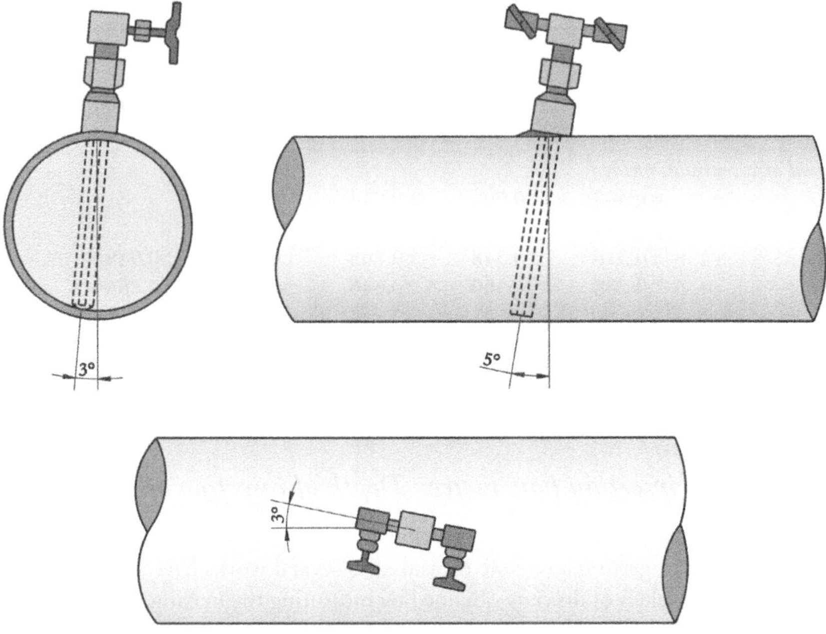

FIGURE 3.33. ADMISSIBLE ERRORS BY A MANUFACTURER IN THE DIRECTION OF
INSERTION OF A PROBE

meter, where to a certain extent, the velocity in all points of the section is considered
(Shercliff's weighting function) the relationship between the average velocity and the
measured velocity is very sensitive to the velocity distribution and the quality of the
velocity profile.

3.3.5. Advantages and disadvantages

Insertion flow meters are widely used by water supply companies to carry out on-site
verifications of their meters. Although the real accuracy is not very high, they can be
very convenient and present many advantages:

Advantages	Disadvantages
• Low cost, especially for large diameters. The probe's cost is almost independent of the diameter of the pipe. • Some technologies can be used in water with suspended solids with no risk of deteriorating the device. They can be used to control groundwater extraction. • A single insertion probe can be used in several points of the distribution network. • Small head losses.	• Low reliability in the measurements compared to other technologies. • Extremely sensitive to the quality of the velocity profile. The required length of straight pipe upstream from the meter may be over 50D. • Need to drill a hole in the pipe. • Narrow range of measurement. Usually 1:10 or lower.

4

Remote meter reading

4.1. INTRODUCTION

The remote reading of a meter allows gathering the data provided by the device without the need of physically accessing it. The process which is done automatically allows for a higher reading frequency and consequently provides detailed information on how the water is consumed. Remote reading, when properly managed, can improve the understanding of the system behaviour and the billing process.

4.2. TRADITIONAL METER READING

Water meters are traditionally required by construction to be visually read and therefore imply that a member of the utility's staff needs to access the meter periodically and obtain the value of the metered volume. This method presents several inconveniences:

- *Cost*: The individual reading of every single meter is a costly process and intensely time consuming. That is the reason why, although there are plenty of advantages associated to high-frequency readings of the meters, utilities have traditionally adopted a compromise solution of reading at relatively long intervals (1 month being the shortest one, but often reading 6 or 4 times a year). Billing does not always correlate to readings and some utilities estimate consumptions for some of their billing periods, being unable to perform meter readings more often. This can even affect the tariffs structure, as block tariffs require that readings are performed periodically within very specific dates.

- *Time*: Directly related to the previous point, the time necessary to perform all readings is a source of additional problems. The interval lapsing between readings is usually too long (even when reduced to 2 months) to adequately manage parts of the system (detailed knowledge of the demand, resource planning, etc.). A long reading interval also presents difficulties to adopt a complex tariff, or at least an adjusted one. For instance, if billing takes place every 3 months (which is not unusual in small utilities) it may prevent the use of a seasonal (winter/summer) tariff structure that is precise enough to be effective.

- *Errors*: Reading meters and creating a record for the measure represent a monotonous and repetitive task, and above all, a manual one. It is therefore frequent to find errors in recorded meter readings (such as errors in the recorded figure, in the assignation of the reading to a user or even in the updating of the commercial information system). Some other times, the errors are somehow voluntary and originate in the estimation of the consumption by the reader when the meter is not reachable or simply to save time ("curb" readings).

- *Accessibility*: Water meters' location can often be a problem in itself. Meters can be located inside or outside the building or household, and be mounted individually or in groups. Regardless of the location, the intent of providing certain protection to the meters usually derives in problems of physical accessibility to the devices for reading purposes. This means that unless a meter can be read from public grounds, the user will always have means to prevent the reading of the device.

- *Readability*: The location of the meter is also a key factor in the readability of the dial. Under some circumstances the meter is difficult to read, either because the visual line is blocked or because fogging in the dial hinders a correct reading. If the meter is rotated to facilitate reading, this may substantially increase the error of the instrument and affect billing.

It is then quite obvious that traditional meter reading represents a slow, costly and unreliable process. However, it is still in practice, and will be for a long time. After all, until now, efficiency and efficacy were not important in this process since the cost and price of water have remained relatively low. However, circumstances are now changing. The value of water is now being evaluated as a whole, from source to tap and back to the source, and this will increase prices and costs in the future. Additionally, and as stressed in this book, meters are the main source and the first link in the chain of information on the water consumption by the users. They also represent the main (and almost only) billing tool in the utility. Water meters should not only stay in good operating conditions, but their readings should also be reliable and accurate. A flawless meter can provide bad data if it is not read properly.

4.3. REMOTE METER READING

The number of configuration possibilities that are available nowadays for remote reading systems is quite high. The fact that this is a field in constant development only adds further alternatives.

It is therefore quite usual to find in a system several mixed configurations depending on the particular needs of every location. The dynamics of remote reading systems require all the components to be compatible with other communication devices and capable of being expanded and upgraded with time. As a matter of fact, manufacturers often recommend starting the implementation in pilot projects and extending the covered areas gradually, as the previous stages get consolidated. Under these circumstances it is quite common that dissimilar systems are used for different zones depending on the needs, the circumstances and the available technology. Figure 4.1 shows, for instance, a combination of systems where M-Bus (meter bus) and radio

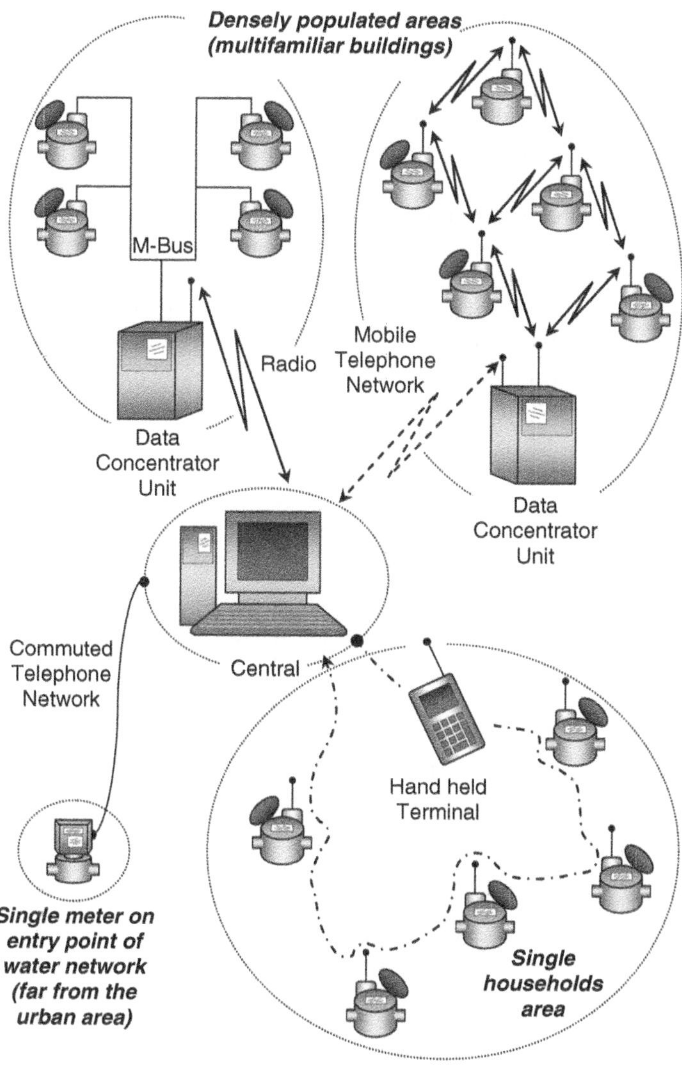

FIGURE 4.1. COMBINED COMMUNICATION SYSTEMS IN A UTILITY

transmission are used (through data concentrator units, DCUs) for densely populated areas, while other alternatives are used for detached households (such as meters with radio transmitters for remote reading with hand-held terminals, HHT), and some other technologies are employed, in turn, for communication between DCUs and the central.

4.4. DATA HANDLING STAGES

Regardless of the specific characteristics of the available remote reading technologies, data transmission follows a similar sequence in every case. Figure 4.1 has already shown the main stages of that sequence. Specifically:

1. The totalizer is the device placed inside the meter in charge of physically registering the flow of water and showing its visual reading.
2. In order to transform the reading of the totalizer into an electrical signal "understandable" and transmittable by the rest of the system, the participation of some sort of transducer is necessary. Consequently, while the meters with an electronic totalizer are capable of performing both functions by themselves (providing a reading and sending the signal) the mechanical meters need an external transducer to generate an electrical signal for the communications system (Figures 4.2 and 4.3). Such transducer is known as the meter interface unit (MIU). Two are the components of the MIU in most cases:
   - A device capable of "reading" the volume registered by the mechanical totalizer and converting it into an electrical signal ready to be transmitted by the system.

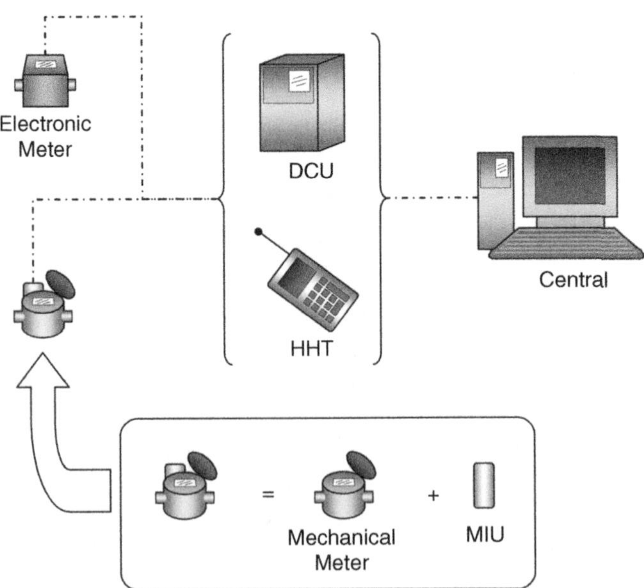

FIGURE 4.2. **DATA TRANSMISSION FROM A MECHANICAL AND AN ELECTRONIC METER**

The presence of this device is completely necessary, and this role is usually played by a pulse emitter.

- Another device which is in charge of transmitting the signal through the system. Depending on whether the transmission is through a wire or wireless, this device will have a different nature. If the system is so simple that the pulses are stored in a data-logger, this second device becomes unnecessary.

3. The data transmitted from the MIU or the electronic meter are received by an intermediate device which function is temporarily store the data from a set of meters until they are transmitted to the central data management system. Figure 4.1 also shows the different alternatives for this device, which may be fixed or mobile.

- When the device is fixed is usually called a data concentrator unit (DCU).
- When the device is portable, it is usually a hand-held terminal (HHT), and the data are downloaded directly onto the computers of the central facility.

4.4.1. The totalizer

The totalizer is the element in charge of integrating the volume registered by the meter through time. The integration is possible due to the fact that the number of revolutions of the sensing element depends on the circulated volume almost independently of the circulating flowrate (within the operating range).

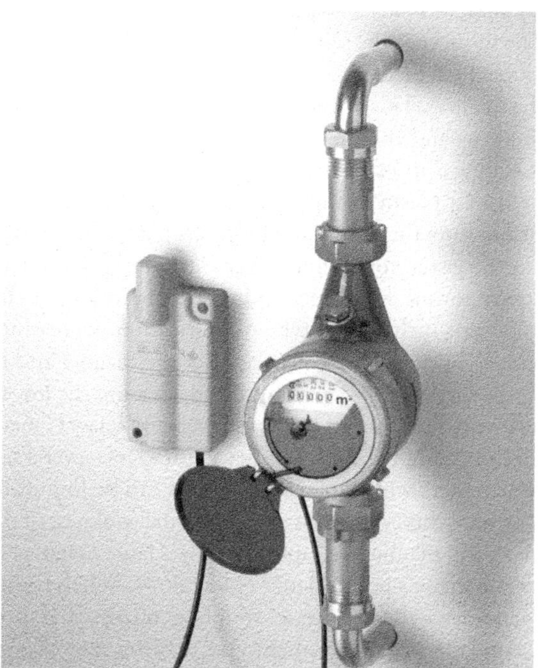

FIGURE 4.3. MECHANICAL METER WITH AN MIU FOR RADIO TRANSMISSION
(COURTESY OF ELSTER)

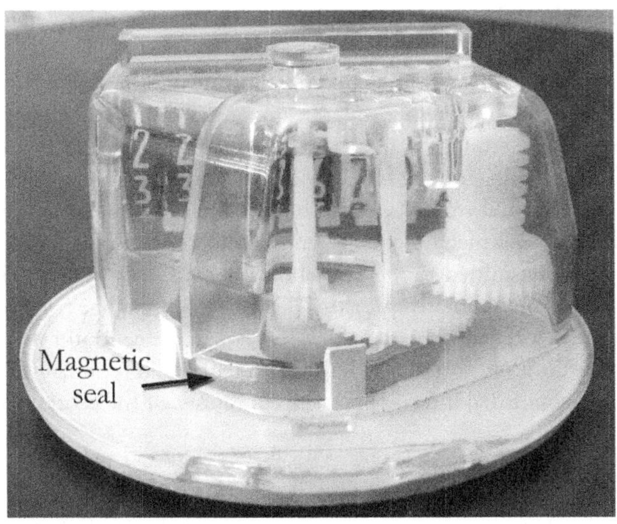

FIGURE 4.4.   EXTRA-DRY TOTALIZER

Totalizers can be classified according to the degree of contact of the gears with the fluid: extra-dry dial, dry dial or wet dial meters.

In the *extra-dry totalizer* meters, no gear or roller is ever in direct contact with the circulating water. The motion of the sensing element (turbine, piston, disc, etc.) is transmitted by two coupled magnets, one of them located on top of it and the other one in the first gear of the totalizer. This is called magnetic transmission and is commonly used in single jet, multiple jet and volumetric meters of small calibre, and also in Woltmann meters. Figure 4.4 shows an extra-dry totalizer, with all the gears and display elements totally isolated in a plastic compartment. The picture also shows an anti-fraud metallic ring that prevents the manipulation by means of an external magnetic field.

The friction in the gears, rollers and other elements which integrate the totalizer generates a resistance torque which determines the metrological class of the meters. Extra-dry meters do not include any type of lubrication, increasing friction and the error at low flowrates. Furthermore, the weight of the magnet and its position in the first gear increase the inertia of the system, thus decreasing the low flow sensitivity of the meter. This prevented the manufacturing of velocity Class C meters with an extra-dry totalizer until the end of the 20th century. Nowadays, several single jet and multijet meters Class C meters with extra-dry totalizers can be found in the market.

A possible problem of this type of transmission is the weakness of the magnetic coupling. Obviously, a strong magnetic coupling requires large magnets which, because of their weight, decrease meter's performance at low flows. When trying to optimize the weight, in some models, the magnetic coupling obtained is not strong enough at high flowrates, and the sensing element and totalizer may slip generating undermetering errors.

In *dry totalizers*, the reduction gears (the first ones in contact with the turbine) are in direct contact with the water while rollers and some gears are isolated from it.

FIGURE 4.5. DRY TOTALIZER WITH ACCUMULATED DIRT

Some parts of the mechanism are therefore exposed to foreign bodies and substances that may alter in the medium term the metrological performance. As a consequence, dry meters are not recommended for hard or dirty waters. Figure 4.5 shows a dry totalizer with accumulated dirt in the gears.

The advantage of this type of totalizer with respect to the extra-dry one is the reduction of friction and inertia that is obtained. For manufacturers it is easier to design a meter with a low flow sensitivity using this configuration. The coupling between the sensing element and the reduction gear is mechanical. However, the dry and wet parts of the totalizer are usually coupled magnetically. The influence of the weight of the magnet is reduced by the de-multiplication of the gear coupling with the totalizer.

Wet totalizers have all the gears completely immersed in water and, in occasions, some other fluid. These totalizers are continuously lubricated and have considerably less friction, thus increasing their sensitivity at low flowrates. As a matter of fact, most single jet and multijet Class C meters are equipped with wet totalizers. In some constructions of wet totalizers, rollers and some gears are protected from water in an isolated capsule (Figure 4.6). This capsule is filled with a lubricating fluid (similar to glycerine). This type of totalizer may use mechanical transmission from the turbine to the rollers.

Meters with wet totalizers are recommended in those installations where fogging may appear, for instance, those in which the temperature of the water is low (with respect ambient temperature) and can present sudden changes, such as those located outside at ground level. The obvious inconvenience is that these meters are extremely sensitive to any impurities in water, and are not recommendable in utilities with a high number of repairs in the network or with hard or loaded water. In these cases, wet totalizer meters often cease to work shortly after their installation.

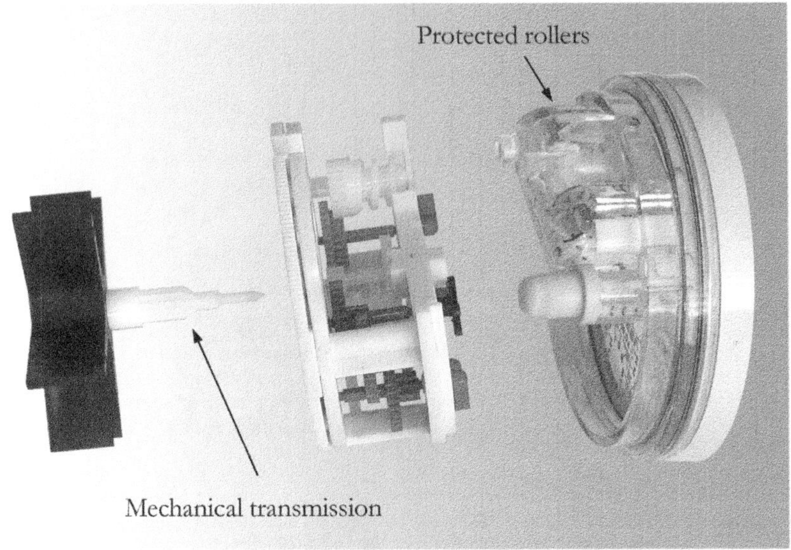

FIGURE 4.6.   **WET TOTALIZER, GEARS AND TURBINE**

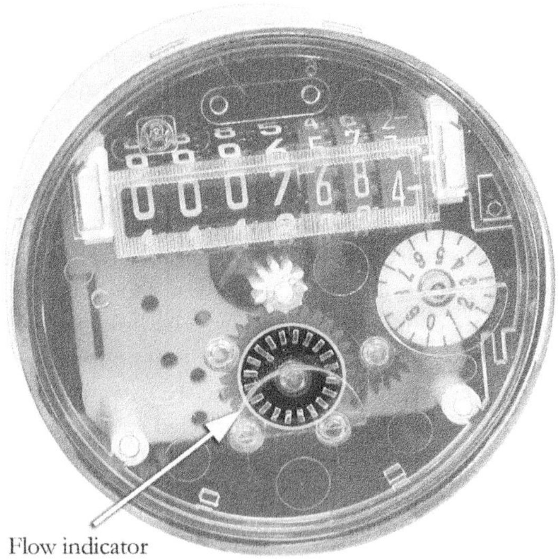

FIGURE 4.7.   **FLOW INDICATOR IN A EXTRA-DRY TOTALIZER**

Finally, and independent of the specific characteristics of the totalizer, there is an additional device that may be incorporated to it, which is a flow indicator. Household leaks are a key factor in the selection of a meter type and in the determination of its lifespan. The flow indicator (Figure 4.7) allows detecting the rotation of the sensing element at very flowrates (for there is little or no multiplication between this element

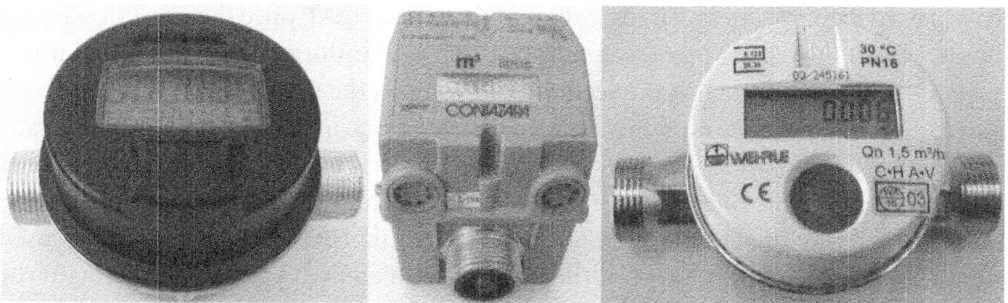

FIGURE 4.8. METERS WITH AN ELECTRONIC TOTALIZER

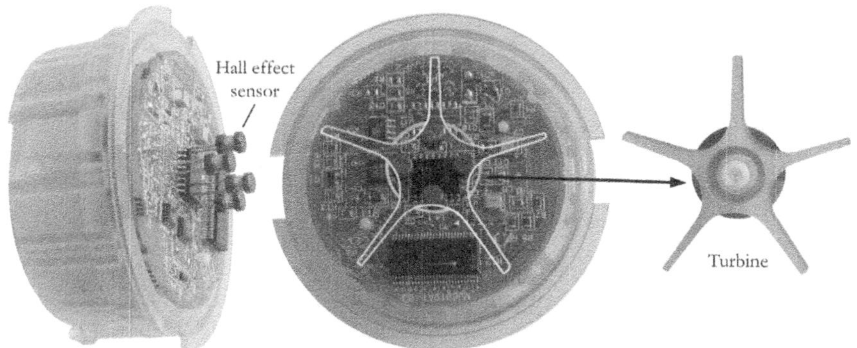

FIGURE 4.9. HALL SENSORS IN AN ELECTRONIC METER

and the turbine). This is useful to verify the existence of leaks in the household installations, or even better, to determine whether an existing leak is detected by the meter. The flow indicator also allows testing the meter at low flowrates in a reduced time, minimizing the costs.

Another classification divides totalizers into mechanical and electronic. All totalizers presented until now are mechanical devices, for the integration of the flowrate is carried out through gears. However, this integration can also be achieved by using a Hall effect cell that detects the motion of the sensing element and accumulates the cycles by means of electronic circuits instead of gears (Figure 4.9).

A Hall effect cell is a sensor capable of detecting the variations in the magnetic flow. In velocity water meters, this magnetic field can be generated by a magnet placed on the upper part of the turbine. The rotation of the turbine generates a shift between the positive and negative poles, and as a consequence, variations in the magnetic field.

The sensor generates a switch in each change of the magnetic field. Consequently, if the magnet has two pairs of poles, four switches are generated in each turn of the turbine. In this case, the volume of water associated to a quarter of a turn of the turbine would be the smallest readable resolution of such meter. In other cases, each blade of the turbine is magnetized with a different polarity, and the sensor is able to count the number of changes of polarity (and therefore the number of blades that have passed through the spot). Regardless of the system used, once the signals generated by the sensor have been processed, it is possible to transform them into useful information for the utility, such as the accumulated volume, the time that the meter has been active/inactive, the minimum detected flowrate, the maximum consumption flowrate, the demand pattern of the user, etc. It is quite clear that the future of water meters lies within the electronic models considering all the advantages involved.

However, one of the main reasons why electronic totalizers are not widely used nowadays is because of the cost of the battery (with enough capacity to maintain the meter at work for at least 10 years) which makes them much more expensive than the mechanical devices. The excess cost is only justified when additional features are included in the meter totalizer.

4.4.2. Pulse emitters

Pulse emitters are the devices used to convert the reading of a mechanical totalizer to an electric signal interpretable by an electronic device.

There are several types of pulse emitters. The most common ones are Reed, optical and inductive emitters. In any case, for a pulse emitter to work some elements have to be added to the totalizer:

- *A mobile component*: This component is permanently coupled with the gears of the totalizer, and as a consequence it is associated with a fixed volume. The value of this volume depends on the gear that "carries" this mobile element. Often, the mobile component is incorporated at the factory inside the meter, and performs its function whether it is used or not.
- *A fixed component*: This component uses to be an external one. It may be attached or not to the meter, and is the one commonly known as "pulse emitter". One of the functions of this element is to detect the passing of the mobile component and emit, through a cable or other mean, an electric or electronic signal which is associated to a fixed volume of water.

In *Reed pulse emitters* (Figure 4.10) the mobile part is a magnet which rotates with the gears. Next to the magnet, a capsule is placed with two metallic plates. The rotation of the magnet attracts or repels one of the plates closing the contact, which can be detected by the appropriate equipment.

The following table summarizes the main features of Reed pulse emitters.

Advantages	Disadvantages
- No external electric supply is needed. - Signal has no polarity. - Cable length up to 1 km depending on the impedance of the line. - Low cost.	- Low working frequency (up to 1 Hz). - The plates can bounce back, generating artificial pulses. - They are sensitive to electromagnetic interferences, affecting their reliability. - The flow direction cannot be detected (unless several switches are used). - The emitters are quite fragile.

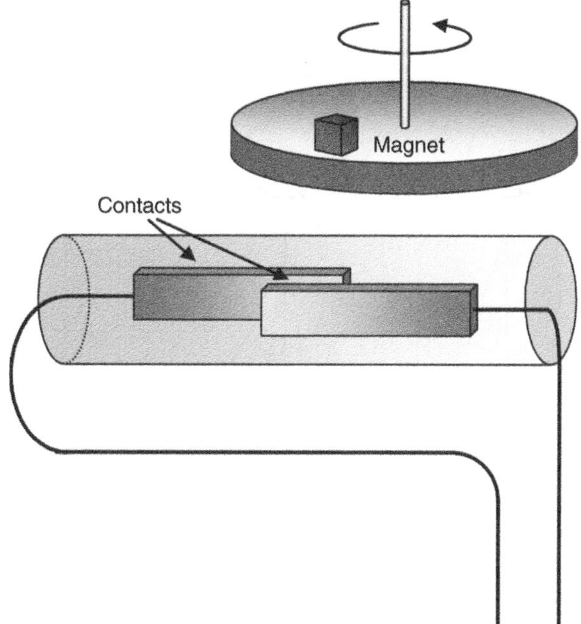

FIGURE 4.10. OPERATING PRINCIPLE OF A REED PULSE EMITTER

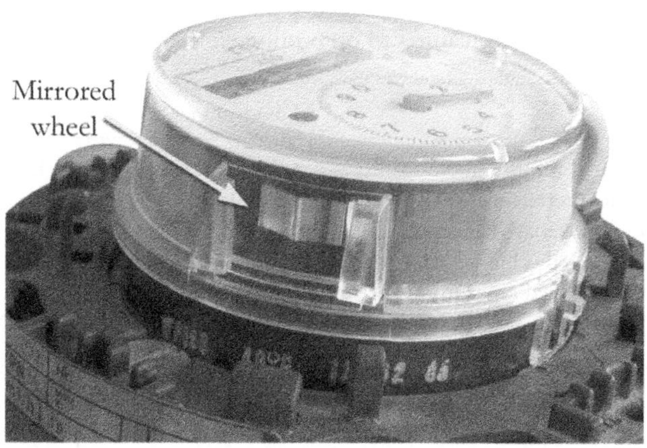

FIGURE 4.11. MIRRORED WHEEL ON AN OPTO PULSE EMITTER

Opto or optical emitters (Figure 4.11) are more complex and present certain additional advantages. The fixed component consists of a light emitter and optical sensors, while the mobile component is a mirrored wheel that is able to reflect the light systematically in certain directions as it rotates.

The light emitter generates a beam of light which incides on the wheel in a fixed direction (Figure 4.12). The shape of the wheel helps to reflect the light beam, which changes its direction according to the rotation. The order in which the sensors detect the

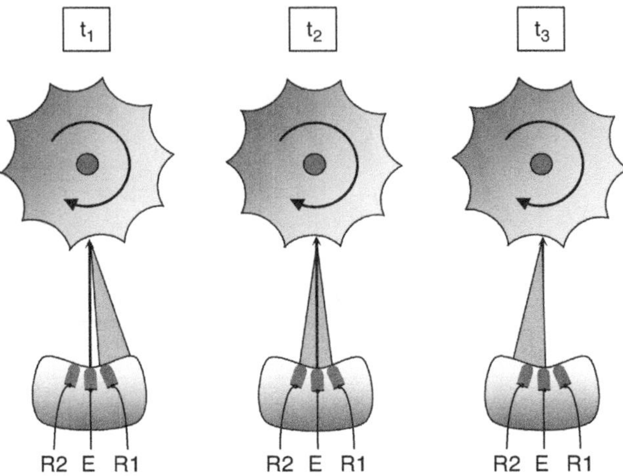

FIGURE 4.12. OPERATING SCHEME OF AN OPTO PULSE EMITTER

light provides information on the direction of rotation of the wheel. The water volume passed is associated to the number of times the sensors detect the emitted light. An electronic system processes both variables and emits the corresponding pulses accordingly through the signal cables or other means. Some other configurations of optical emitters simply use as a mobile component a partially mirrored wheel that generates a signal every time that a change between a mirrored and a non-mirrored section is detected.

Summarizing, the main features of opto pulse emitters:

Advantages	Disadvantages
• The signal emitted is more stable against electromagnetic interferences.	• External electric supply is needed.
• High working frequency. The amount of information supplied is high.	• Available only for medium and large calibre meters.
• More robust than Reed type emitters.	• High acquisition cost.

Last, but not least, are the *inductive type emitters*. They are the most reliable and present the highest number of alternatives among the available technologies nowadays. In this type of emitters, the mobile component is a simple metallic plate coupled to one of the dials of the meter (Figure 4.13). The fixed component is formed by an inductive element (coil) which is fed by a small electric current.

While registering volumes, the dial will rotate and the metallic plate will periodically go through the magnetic field created around the coil, thus creating an alteration in the electric current (Figure 4.14). These alterations allow identifying the position of the metallic plate and therefore the number of turns of the dial. For this reason the volume associated with each pulse depends on the dial to which the metallic plate is attached.

The inductive type emitters can also provide additional information like flow direction and cable cut detection.

FIGURE 4.13. INDUCTIVE TYPE EMITTER

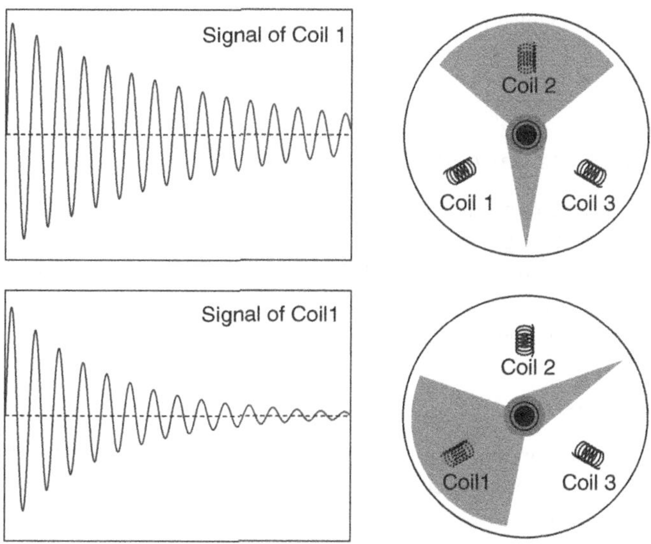

FIGURE 4.14. ALTERATION IN THE ELECTRICAL CURRENT CIRCULATING THROUGH THE COILS IN AN INDUCTIVE TYPE PULSE EMITTER (COURTESY OF ACTARIS)

Summarizing the main features of inductive pulse emitters:

Advantages	Disadvantages
• No external electric supply is needed, although they operate on batteries inserted on the fix component. • May work at high frequencies. • Provide additional information to the accumulated volume. • Mechanically more robust than Reed type meters, with a similar price. • The same type of standardized pulse emitter can be used in all compatible meters.	• Operating life limited by battery power (usually more than 10 years).

4.4.3. The data concentrator unit (DCU)

A DCU is, in essence, a temporal data storing device that transmits the collected information later. The purpose of the DCU is to gather the readings of a set of meters (that may be quite large) and later send them to the central data processing unit of the utility, or even a hand-held terminal or a laptop computer. The DCUs may transmit the data both by cable and radio (Figure 4.15).

Cable transmission

The simplest setup for the communication between a meter and the DCU is the independent connection of each meter with a cable. This parallel configuration is shown in Figure 4.16.

Although this is the most obvious way to connect the system, it is neither the most practical nor the most efficient, for long cables usually become impractical and ravelled. Furthermore, in most places the cost of laying long cables makes this type of connection unaffordable.

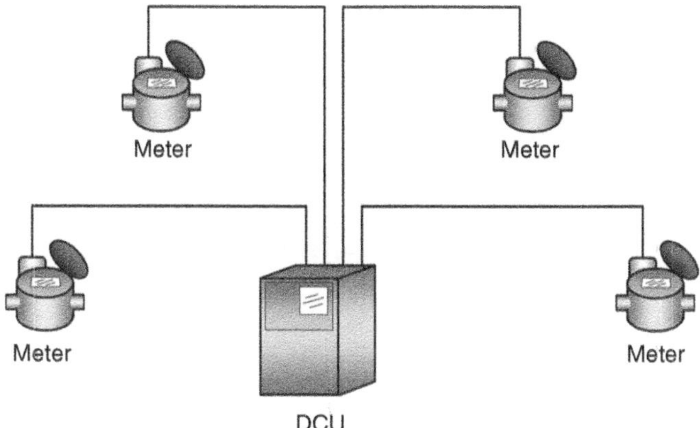

FIGURE 4.16.   METERS–DCU PARALLEL CABLE CONNECTION

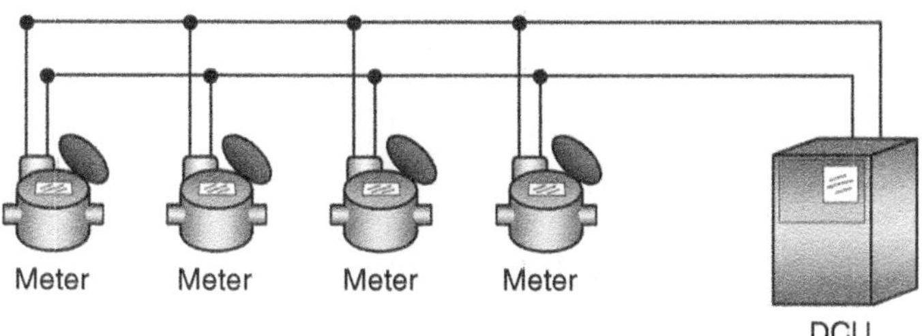

FIGURE 4.17.   SCHEMATIC OF METERS IN A BUS CONNECTION

An improved cable connection would be the bus connection. Figure 4.17 shows how meters are connected to a data bus which consists of two cables which deliver the information to the DCU (Figure 4.22 shows an actual installation of meters in a bus connection). In this setup, the data bus actually goes through the meter casing, thus giving the impression of a false serial connection.

A bus connection is obviously more efficient in the use of cable, and it is easier to install. However, there are some associated problems that need solution. Firstly, since all data from the meters must travel on the same cable (data bus) a protocol is needed to allow queries from the DCU to each individual meter, while discriminating adequately all the information. This also implies that every meter needs to be equipped with some sort of "intelligence" that will deliver the data in due time, needing a small electrical supply. Last, it is also important to point out that this type of connection may be less robust than a parallel connection, since any problem with the cable connecting the meters will affect the signal of all meters behind that point.

There are several protocols available for bus connections. The M-Bus (meter bus) (Figure 4.18) could be considered a standard, for it is the most used and is included in the European Standard EN 1434-3. Systems based in this protocol can connect up to 250 meters in a single DCU. The system is very versatile and can also integrate meters from different manufacturers, or even, from other utilities such as gas and electricity.

FIGURE 4.18.   PULSE EMITTER WITH M-BUS COMMUNICATION APPLICABILITIES FOR A
WATER METER (COURTESY OF ACTARIS)

FIGURE 4.19.   PULSE EMITTER WITH RADIO COMMUNICATION APPLICABILITIES FOR A
WATER METER (COURTESY OF ACTARIS)

Radio transmission

The alternative to cable communication is the wireless transmission. Usually, this wireless connection between meter and DCU is achieved by radio waves, and consequently a small radio module is needed in every meter (Figure 4.19). Thus, the meter only needs to be close enough to the DCU in order to allow the data transmission.

The radio transmission presents an important number of advantages, but also some disadvantages. The absence of cables, on one hand, makes installation easier and cheaper, being able to cover larger areas and having longer distances between meters. On the other hand, radio transmissions will never be as reliable as cable connections as they are subject to interferences that may be caused by metallic elements, electronic devices, physical obstacles or even the weather conditions.

A radio system can also adopt several configurations. Once again, a parallel configuration is the most intuitive one (Figure 4.20) with the individual transmission of

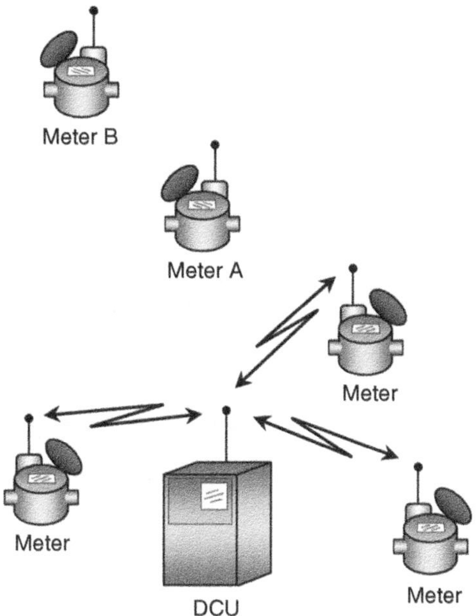

FIGURE 4.20. SCHEMATIC OF A WIRELESS METER CONNECTION

the information travelling from each meter to the DCU. However, this configuration requires all meters to be close enough to the DCU to allow a correct transmission of the data. If meters are situated outside the DCU transmission range they will not be able to transmit the data (for instance, meters A and B in Figure 4.20).

To solve this problem, the new radio systems allow the connection of all meters in a mesh in order to get the signal to the final DCU through the network nodes. These nodes are elements which provide consistency to the network, and this role can be played by specific repeaters whose only function is to re-transmit the signal (meters on the upper-right side of Figure 4.21), or even by radio modules attached to the meters capable of performing this function (meters on the left side of Figure 4.21).

The mesh configuration is useful to solve the problem of distance between meters and the DCU. Additionally, it is also very versatile. For instance, the system itself will configure a new signal routing if the communication through the usual route is interrupted. This feature increases the robustness of data transmission making it less sensitive to interferences. Some systems will even reconfigure themselves periodically searching for the optimum transmission routes.

The radio transmissions capabilities are obviously dependent on the specific models and technologies available in the market, which are constantly evolving. Usually, the frequencies used range from 400 to 900 MHz and the range of transmission can be estimated in 20–50 m inside buildings and 100 m outside. The batteries of the devices usually last over 10 years.

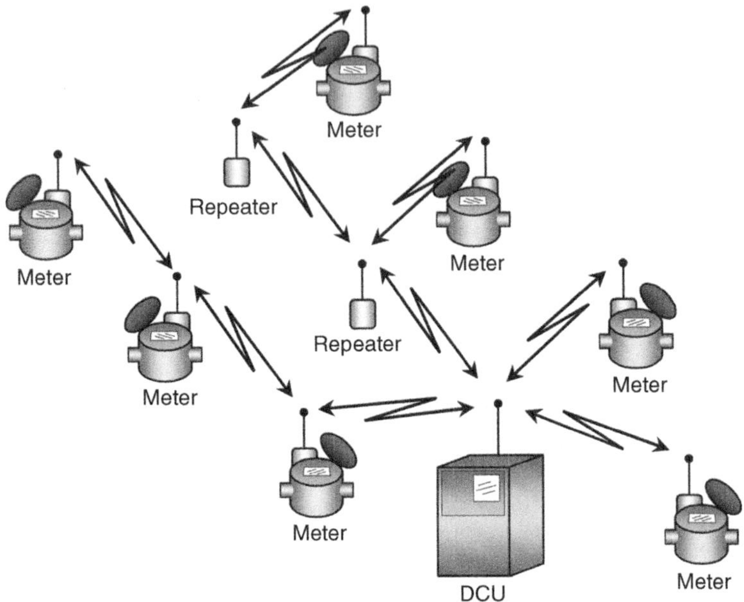

FIGURE 4.21.   SCHEMATIC OF A WIRELESS METER CONNECTION WITH SIGNAL REPEATERS

Directionality of the transmission

An important concept of transmission between the meter and the DCU is the directionality of the communications. In both cable and wireless transmissions the communications may take place in one direction (unidirectional) or two directions (bidirectional).

In unidirectional systems, the information travels only from the meter to the DCU, without any kind of feedback. Once the meter is installed, the registered volumes are sent periodically regardless of whether there is remote reading equipment receiving the data or not. This type of transmission is typical of simple systems that consist of a pulse emitter and the transmission capability (via radio or cable).

In bidirectional systems, the information can be transmitted in both directions, from the meter to the DCU and vice versa. During the normal operation of these systems, the meter only transmits data after the DCU or the HHT have issued a request. The meter remains in stand-by mode until it is queried by the DCU. It is only then, when the meter is activated, that it sends the information and if the process is completed successfully, it returns to stand-by mode until the next query. M-Bus standard uses this sort of communication. Most radio systems also use bidirectional transmissions.

It is obvious that bidirectional communication is more complex and requires a protocol to allow queries and replies between the meter and the DCU, as well as a certain process capacity from the meter or the pulse emitter. However, there are also some significant advantages. The system is much more flexible and data on water consumption can be obtained in real time when required.

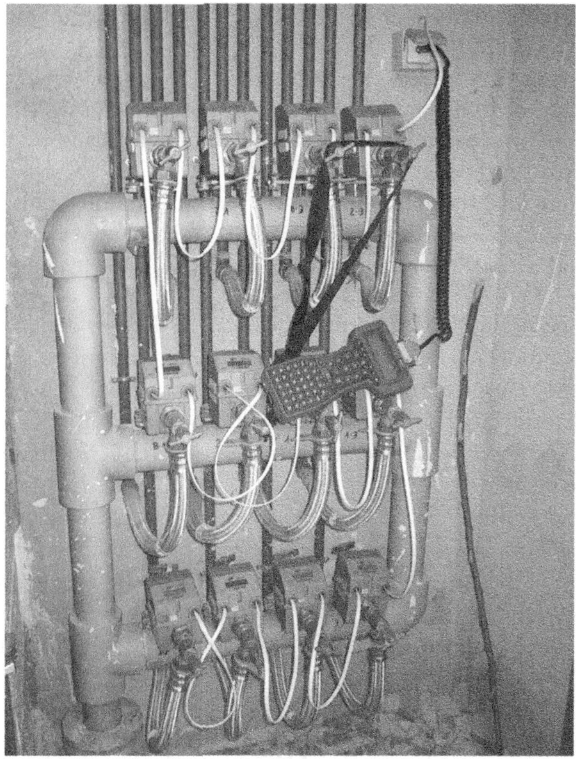

FIGURE 4.22. HHT CABLE CONNECTION (COURTESY OF CONTAZARA)

4.4.4. The hand-held terminal (HHT)

An HHT is a small electronic device similar to a PDA, though slightly larger and more robust, that is used to collect the data directly from the meters (Figure 4.22).

Depending on the type of communication, the hand-held terminal (HHT) needs to be within a close range from the meter to be read (radio), or even physically connected to it, in order to download all the data (cable). The data is stored in the HHT internal memory. Afterwards, the HHT is returned to the central offices where the information is downloaded into the commercial system. In this way, the HHT works like a sort of mobile DCU.

The standard working procedure with HHT presents many similarities to traditional meter reading, for it is necessary to visit the different locations of the meters (like in traditional meter reading routes) and the information is afterwards delivered to the central offices. It is therefore questionable whether HHT constitutes a remote reading system, or simply an automatic reading system in which the human participation is still needed in the reading process (though not to read the meters, but to carry the HHT to different locations). Regardless of these considerations, the multiple advantages of HHT data collection have made it extremely popular, and these minor terminological aspects can be ignored. Some specific advantages of using an HHT are the following:

- Data are collected from a group of meters. In a single stop the data from all individual meters in the surroundings can be collected, saving time. The reading routes are reduced in length and number.
- The procedure for data collection does not depend on human intervention. This saves time (data is downloaded much faster than if the meter readings have to be typed in) and eliminates errors in the transcription of the readings.

Data transmission to the HHT from the meters can be done by cable and radio. Cable is less flexible but more reliable, and needs a specific connection for the HHT next to the meters. Radio transmissions can be affected by interferences, but the connection only requires that the HHT is somewhere near the meters in order to collect the data. In that case, collecting the data becomes a matter of simply strolling by the different locations of the meters.

One further improvement is still possible: collecting the data via radio from a vehicle. With a quick radio transmission, and no need to visit the buildings, the single process demanding most of the data collection time is taking the HHT from one DCU to the next one. Consequently, by installing the appropriate equipment (a portable computer with a radio modem instead of an HHT) in a vehicle the travelling time will be reduced significantly.

4.4.5. Transmission from the DCUs to the central information system

Once all readings from a group of meters have been gathered by a DCU, the information should now be sent to the central information systems of the utility. This second stage of data transmission is different from the first one in several aspects:

- The volume of data to be transmitted is much larger, for the data of all meters (up to 500 depending on the system) need to be sent.
- The distance is also much larger. DCUs have to be located near the meters, but the DCUs themselves can be very far from the central data collection facility.

For these reasons, there are more available means and technologies for this second stage of data transmission. Figure 4.23 shows the available choices.

Radio

Although not as popular as other systems for this second stage, radio frequency is also used to transmit information from the DCUs to central systems. The only requirement is to equip DCUs with the adequate radio modems, setting up a second radio network. This network is more powerful and allows the transmission of data from all DCUs to the central information systems.

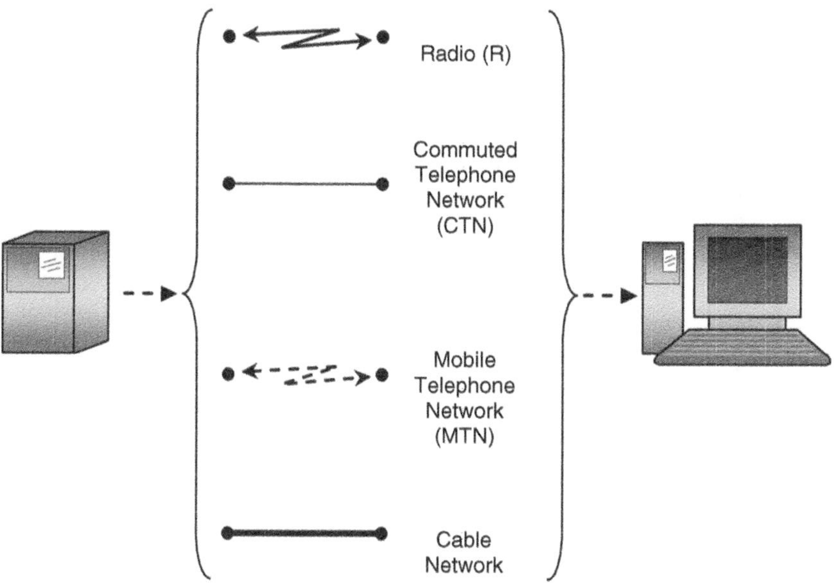

FIGURE 4.23. DCU–CENTRAL TRANSMISSION TECHNOLOGIES

Commuted telephone network

Another alternative for long distance data transmission is to use the telephone networks present in urban areas. The traditional approach would be the commuted telephone network (CTN), which requires the use of an analogue modem connected to the DCU and a physical telephone line at the installation. With this simple equipment and the adequate electrical supply for the modem, the DCU will transmit the data with standard local telephone calls. The DCU may also be configured to make the calls at the hours when telephone rates are cheapest.

Mobile telephone network

An obvious alternative to the previous one is to use the mobile telephone network (MTN; for instance, in many parts of the world the GSM network). With the installation of a compatible modem, everything said for the CTN can be also stated for MTN (Figure 4.24).

Given the characteristics of GSM networks, data transmission can be achieved in two different ways:

1. By means of encoded telephone calls (similarly to CTN).
2. Using the short messaging service (SMS). This alternative presents several advantages, such as reduced costs, a higher reliability (the message is only sent when the connection is possible), a better coverage (the signal strength needed for calls is higher) and the fact that in case of failure in the communications, the message remains stored by the operator avoiding the loss of communications. The only

FIGURE 4.24.   DATA TRANSMISSION THROUGH A MOBILE PHONE NETWORK
(COURTESY OF CONTAZARA)

disadvantage may be that the information transmission is slightly delayed (the information is not delivered in true real time).

Cable network

The use of urban cable networks for data transmission is not very extended yet. This alternative makes use of the existing cable networks (which usually provide television, Internet and telephone signals) to transmit the data to the central information systems. This alternative would require the installation of a cable modem and the adequate software at the DCU to make use of the Ethernet standard.

4.5.   INFORMATION PROVIDED BY A REMOTE READING SYSTEM

One of the advantages of remote reading systems is that, in some cases, the meters are able to provide much more detailed information than the simple accumulated volume. This information may be used to improve network management and reduce water loss.
    Generally speaking, remote reading systems can provide the following data:

- Total volume accumulated by the meter, date and time. It allows the calculations of water balances to be improved, since information on water consumption for short intervals is available. It also allows efficiently setting block tariffs avoiding user complaints.

- Total working time and total stopped time. Helps detecting excessive consumptions due to leaks or failures inside user's facilities.
- Number of startups.
- Consumption histogram. Registered volume for each flowrate interval in a predetermined series. This information is specially useful for sizing meters in large consumers. It provides the usual range of operating flowrates for the meter.
- Alarms (including date and time for each one of them):
  - Excessive consumption (possible burst).
  - Average consumption extended for too long (possible open tap or open toilet tank).
  - Continuous low consumption (possible leak).
  - Meter malfunctioning (possible failure or manipulation).
  - Low battery.
- Stratification of the consumption by the hour. In this way, it is possible to favour the consumption in off-peak hours and laminate the demand curve, reducing peak consumption.

It is obvious that if this data is delivered on a continuous basis, the range of management options to improve the system's performance is increased significantly. Some manufacturers have even developed an Intranet application that allows displaying and managing all these data in a web environment. The different security levels allow that the corresponding information can be accessed by the different stakeholders, i.e. the user will be able to access his/her personal statistics, historical data and real-time consumption figures; and course, all of the data will be available for the corresponding departments of the utility.

In any case, it should be pointed out that a large amount of data does not necessarily mean that useful information is available. A utility setting up a remote meter reading system should also invest in the necessary resources to process and analyse the collected data. Otherwise, it will be difficult to integrate the gathered data in the decision support systems of the utility, and even more difficult to get an adequate return on the investment.

5

Integrated water meter management

5.1. INTRODUCTION

An integrated water meter management approach within a utility comprises all the actions and decisions aimed to increase the reliability and accuracy of the meters and, at the same time, reduce the overall costs of the installed meters. All these actions take place in a cycle which begins the moment a meter model is selected and finishes when the meter is replaced. Between those two moments additional actions may be undertaken to improve the way water volumes are registered and to reduce the investments related to water consumption measurement. This chapter intends to organize some of the information presented in the book and introduces an integrated system to improve metering efficiency in water utilities.

Ideally, each meter in a water utility should be managed individually. The meter should be selected according to the operating conditions, taking into account its costs and the associated marginal revenue (resulting from a different meter choice). The objective in choosing a certain meter model should be to maximize the net annual benefit the meter provides during its life cycle.

Once the meter model has been selected and purchase from the supplier it will then be necessary to carry out quality control procedures at the reception of the meters. The procedures should not be limited to a simple determination of the meters' error, and are necessary to verify that the instruments meet the expected metrological and constructive specifications.

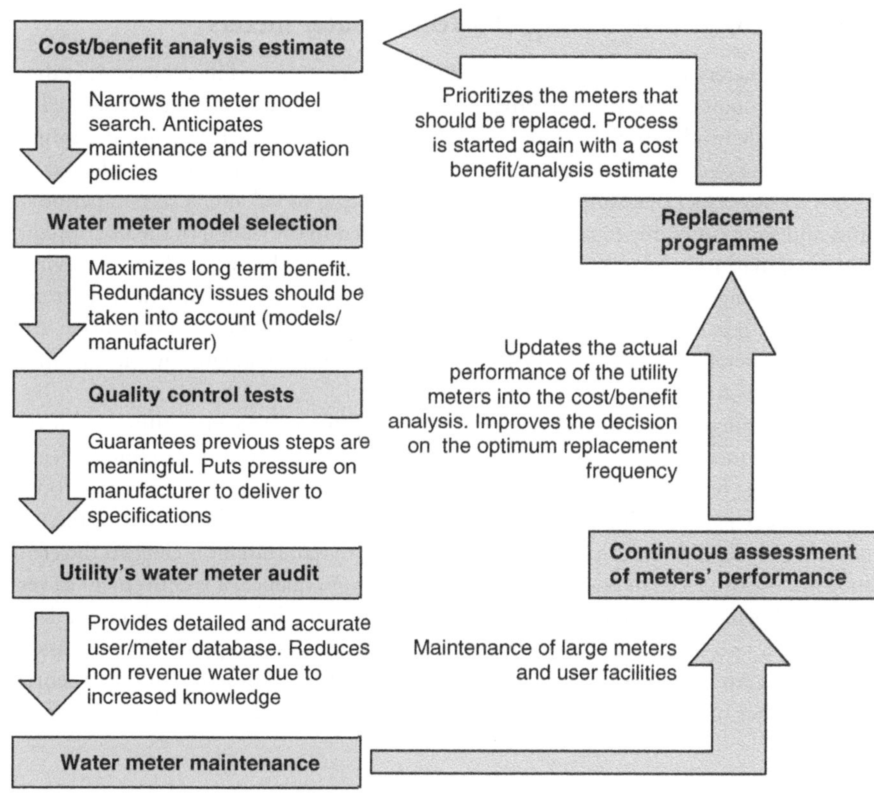

FIGURE 5.1. INTEGRATED METER MANAGEMENT RELATED ACTIVITIES

After the installation of the meters, the management of the meter will comprise the supervision of the installation and the related maintenance and renewal activities.

Finally, one aspect that should not be forgotten has to do with meter reading management. It is easy to understand that even for exceptional meter accuracy, if readings are not handled in an adequate manner, non-revenue volumes will increase significantly.

5.2. SELECTION OF THE METER MODEL

The first obvious task in the management of a meter is to select the appropriate model for each circumstance. This process implies an economic analysis which involves data which are either *assumed* or *estimated* regarding the weighted error decay rate of the meters. These data should be ratified with real information obtained from the field. As a general rule, and in the absence of other restrictions, the chosen meter will be the one that maximizes the annual benefits during its life cycle.

However, there are other factors that need to be taken into account while selecting a meter, such as the quality of water, the frequency and nature of maintenance works in the network (which may affect the presence of suspended solids), etc.

5.2.1.   Initial field testing of selected water meters

Most water meters are complex mechanical instruments and it is difficult to predict their future metrological behaviour in advance. For this reason it is not recommended to install a single model for the entire network. Should such model present functioning problems, the overall registered consumption volumes could be seriously affected.

It is a known fact that some meter models only begin to fail after a certain period of time, and sometimes due to causes not directly related to the design of the instrument. In these situations, the economic losses can easily be much greater than the savings obtained through the selection of a single model for the entire utility. Under these circumstances, it is usually very difficult to determine and allocate responsibilities, and losses are generally assumed in the first place by the utility and later by the customer through the water tariff.

As mentioned in previous chapters, the parameters with greater influence in the behaviour of meters are related to the quality of water, the consumption flowrates and the measuring technology of the meters. For instance, limescale build-up in the body walls of single jet and volumetric meters may, with time, lead to the turbine or the piston being blocked. In multiple jet meters, when the by-pass circuit is clogged the error curve may be displaced to positive error values. As a consequence, it is convenient to verify the real behaviour of a meter model in the field, testing it in real conditions at least for a year or two in a significant sample population of users and previous to a massive installation. An initial testing period of the meters will prevent many of the potential problems that may appear with time depending on the water supply characteristics.

5.2.2.   Economic selection guidelines

In a large number of occasions, meter selection is strictly related to the retail price of the meters or the expected initial error. Such is the case when a utility decides to install Class C meters instead of the current Class B models. However, before making such a decision, it is convenient to evaluate the different alternatives from a comprehensive economic perspective. This analysis is always convenient, even in the absence of real data regarding the accuracy rate of decay of the meters (Chapter 7 shows how to carry out a sensitivity analysis of the accuracy depending on the different parameters considered). Once real data become available from the field and laboratory testing of the meters, the estimated values will be replaced with real ones.

However, and regardless of the economic analysis described in this book, it is convenient to make some additional considerations. Firstly, the price of water may be the variable with a greater weight on the selection of the most adequate type of meter, and it certainly is decisive in determining the replacement period. It has been shown, that the ratio V (as defined in Chapter 7) between the revenue from a meter and its price (including installation) is directly related to the optimum renovation frequency. However, the ratio V also bears a direct relationship with the possibility of upgrading the meter to a higher-quality model.

The consumption pattern is another parameter which greatly affects the type of meter to be used. If leaking plumbing systems are frequent in a utility, it will be necessary to select meters of better accuracy than in a situation where the leakage in indoor facilities is moderate. Additionally, in the first case, the replacement period of the meters, for a given water price, will also be much shorter.

The worst case, from a metrological point of view, appears in situations where buildings are equipped with private storage tanks regulated by proportional flapper valves. The resulting consumption pattern concentrates the majority of used volumes in the lower range of flowrates. Under these circumstances the use of meters which are accurate at low flowrates is a necessity. A common alternative, when the local regulations allow it, is to install meters of lower nominal flowrates, increasing sensitivity at low flowrates without the need of a better metrological class. This is a normal procedure in some countries where the nominal flowrate of the meters for domestic users is reduced from 1.5 to $0.6\,m^3/h$. This will reduce, for instance, the minimum flowrate of a Class B meter from 30 to 12 l/h.

The final factor influencing both, the maximum acceptable price for a meter and the replacement period, is the discount rate selected by the utility for the economic analysis. For this, it should be taken into account that meters cannot be considered a standard investment for they often provide benefits that can hardly be accounted for in economic terms (social, environmental, etc.). As a matter of fact, if the economic assessment is carried out with high-discount rates (above 5%) the results hardly ever advise investing in higher-quality meters. In these cases it is also common to obtain renovation periods of over 20 years, which from a social, metrological and environmental perspective may be inadmissible.

5.3. QUALITY CONTROL

Just like any other measuring device, water meters must go through several quality controls (before they are installed and throughout their life). The aims of these tests are to determine both the quality of the instrument and its compliance with regulations and characteristics, and also to evaluate the capacity of the instruments to adapt to specific field conditions.

5.3.1. Quality control at the reception in the laboratory

Once a meter model has been selected among those available in the market, the quality of the devices received by the utility must be tested. In principle, this should not be a necessary procedure, for all meters are factory tested individually at three different flowrates – minimum, transitional and maximum – according to current standards. However, the possibility of problems during transportation, manipulation or even of manufacturing defects advices independent testing. As a matter of fact, it is possible, and should be considered, that a number of meters in every shipment will not conform with the requirements of their metrological class. The packaging of the meter is a key factor in this process, absorbing the shocks that the meter may suffer during transportation and protecting the device until it reaches the user.

The tests carried out on samples of the acquired meters must follow meticulous test procedures (described in Chapters 8 and 9). These procedures will guarantee that the meters conform to the standards and the tests have adequate uncertainty and repeatability. Following strict testing protocols in the laboratory is essential to avoid erroneous conclusions. For instance, complete draining of the test line, adequate reading of the meters, guaranteeing the stability of the flowrate and adjusting the values of test flowrates to the correct interval (especially at the lower end of the range, where errors evolve quickly). However, for velocity meters, it is possible that even following a strict testing procedure, the errors resulting from the tests differ from those obtained by the manufacturer in the initial verification. In such case, it is convenient to perform a laboratory comparison with the provider to determine the origin of these differences.

The criteria for acceptance or rejection of shipments must be established and agreed with the provider beforehand, for both parties are exposed to certain risks in the operation (for only samples are tested, never the whole shipment). The buyer runs the risk of accepting a defective shipment, while the seller may get a lot of good meters rejected. The agreement on sensible terms for the acceptance and rejection of shipments will usually depend on a balance between the risks assumed by each party.

5.3.2. Quality control on the field

In addition to any tests carried out in the utility's laboratory to determine the quality of the purchased meters, it is convenient to define a protocol to test meters on the field. Such a standardized procedure will allow controlling meters under known working conditions, providing data which are essential for later meter selection procedures. These tests could include the determination of variables such as:

- The meter's capacity to measure leaks and low flowrates from the users' facilities.
- The degree of adaptation of the meter to the water quality, the network operating pressures, humidity, temperature, etc.
- The influence of peak flowrates and demand patterns in the mechanical decay of the meters.

A method often used by technical staff on the field is the serial installation of different meters to allow comparison. This is a useful method and provides comparative information on the meters' performance in identical operation conditions. However, it is important to take a couple of considerations into account.

Firstly, the possible distortion in the consumption pattern of the users. The installation of several meters in series will affect the total pressure losses leading to the users' tap, thus reducing the maximum flowrate available to the user. However, it should also be taken into account that the leakage rates in the building installations will not be affected by this circumstance.

Additionally, it must be considered that this method is actually relying on a sample of users to analyse the behaviour of the meters. As a matter of fact, the differences in performance between the several meter models will greatly depend on the characteristics of the sampled users. Consequently, it is convenient to determine beforehand the consumption pattern for each sampled user. This will allow relating the differences in meter performance to the characteristics of user and installation.

Another alternative method to determine the error curve of the installed meter consists in carrying out field tests. These tests are performed by comparison with a reference measuring device. This procedure, described in detail in Chapter 8, is subject to high uncertainty and could provide confusing results. As a matter of fact, the uncertainty associated with the test is not only dependant on the theoretical metrological performance of the reference device, but also on the actual procedure followed during the test.

5.4. GENERAL RECOMMENDATIONS FOR THE INSTALLATION OF METERS

The installation of any measuring device is a key factor in guaranteeing quality measures. The meters have to be installed in order to guarantee the metrological behaviour, but also the accessibility and protection of the device. As a matter of fact, at the time of installation, the tasks of reading and maintaining the meter should be taken into account. A clear case to be avoided, especially in small calibre meters, is the installation inside the customers' household. In such cases, the presence of the customer is required for any maintenance operation and even for the periodic reading of the meter. In most utilities, the average error of meters which are installed inside buildings is quite higher than the rest. These meters are usually older and lack maintenance.

For this reason, it is extremely useful to standardize the installations requirements depending on the type of meter, diameter, use, etc., in order to guarantee the quality of the measures:

- The majority of water and flow meters are sensitive to some extent to distortions in the velocity profile. Only volumetric meters are free from being affected. This is why all meters need enough length of pipe ahead of the device (depending on the source of distortion of the profile) in order to guarantee an accurate measure. The requirements will obviously depend on the characteristics of each meter, and the manufacturer's recommendations should be taken into account in this respect.
- In mechanical devices with moving parts, the orientation of installation may increase friction forces and reduce the capacity to measure low flows. Some water meters are designed to work in all positions (Chapter 2). In those utilities where the physical space available for the installation of meters is reduced, it is advisable to install meters able to operate in any position. This policy will improve the measurement of low flows on the short term and increase the average life of the devices.
- Some meter models (like positive displacement meters) are extremely sensitive to the presence of foreign particles in the water flow. Under these conditions it is convenient to ensure that water is filtered before entering the metering chamber.
- In utilities where interruptions of the supply are common, or in elevated positions, where trapped air may be found, it is convenient to install air valves upstream from the meter. The objective is to avoid air circulating through the device, which could generate high rotation speeds for the turbine.
- The replacement of any meter will require the corresponding section to be isolated from the rest of the system. The installation kit should always include isolating valves that guarantee water tightness and are reasonably easy to operate in the few times that they will be used.
- Water and flow meters are devices that get older and their performance may evolve with time. The installation must take into account the need of an intake to perform

the on-site calibration of the device with the necessary warranties. Quite often, the budget limitations limit the installation of by-passes and isolation valves, not making possible future calibrations of the device.

- In areas where frost is a problem, the meters should be protected by expansion vases or any other alternative method. Some models actually incorporate frost preventing measures. In areas with high temperatures the device should be protected from direct sunlight to avoid the degradation of the plastics.
- All electrically equipped instruments should be surge protected, especially those caused by lightning.
- Any measuring device is useful as long as it is possible to make an adequate reading of the measure. The water or flow meter should be installed to allow a reliable and quick reading of the measure. If the visual reading of the device is not possible, alternative remote methods should be incorporated to allow reliable, and economical readings of the measure which also imply low electrical consumption.
- As mentioned before, the installation of the device should allow the performance of maintenance tasks, including replacement. It is also recommendable to allocate some extra space for additional instrumentation such as data acquisition equipment, pulse emitters, pressure gauges, etc.

5.5. METERING SYSTEMS AUDITS

The adequate installation of water and flow meters is not only limited to the physical placement of the device on the field. Additionally, it should also comprise several other aspects that need to be checked when auditing the metering system:

- The vulnerability to fraud, either by manipulation of the metering device or of the installation.
- The accordance of the installation conditions to the ones required by the specific meter model and size.
- The performance of the metering device within the limits of its metrological characteristics through time.

Regarding fraud, it is quite obvious that the utility needs to control the different possibilities in which customers are intentionally manipulating the metering system to artificially reduce their bills. Any customer with low or inexistent consumption (compared to the regular values of the activity) should be checked. In order to do this, the meters can incorporate alarm signals that warn the utility's staff of abnormal values in the consumption. The installation should also incorporate ways to check that the meter has not been manipulated and that the operating conditions correspond to the specifications of the actual meter installed.

A checklist can be elaborated to enable the utility's staff to check the possible anomalies in the performance of the meter as well as characterizing the user in accordance to some predefined parameters. The collected information, once it is conveniently stored in appropriate databases, can be used to perform cross checks with the different stored parameters.

Description of the device

- Operating principle/technology.
- Metrological class.
- Manufacturer's data: model, brand, manufacturing year, calibre, nominal flowrate.
- Integrity of the meter (for instance, against manipulation, status of the seals, etc.).
- Possibility of connexion of pulse emitters.

Characteristics of the installation

- Diameter and material of the upstream and downstream pipes.
- Existence, type and diameter of isolation valves.
- Length of straight pipe upstream and downstream of the meter.
- Ambient conditions that may affect the meter or the reading (humidity, dust, etc.).
- Accessibility for reading and installation position (vertical/horizontal/at an angle).
- Presence of filters and flow stabilizers.
- Existence of pressurized intakes for a meter test on-site.

FIGURE 5.2. CHECKLIST FOR ROUTINE CONTROLS OF INSTALLED METERS

Figure 5.2 shows a sample checklist which may be used to perform a routine control at every meter location. This sort of procedure should be specially applied to large customers. Other items in the checklist should be verified before the activation of the service connection or after every meter replacement.

The designed checklist should be comprehensive enough to completely characterize the meter. The staff in charge of the inspection should also verify the integrity of the meter looking for malfunctioning or manipulated meters. Special attention should be paid to the metrological seals that guarantee that the device retains its factory specifications. The manipulations of the totalizer are quite frequent and this element should be carefully inspected. If the meter is capable of mounting pulse emitters, the communications system to be used should be noted. In electronic meters the type of output signal should also be registered: intensity (mA), voltage (V) or frequency (Hz).

It is also important to determine if the meter has been correctly installed. Especially relevant are the installation conditions (diameter and material of the connection pipe, the length of straight pipe upstream and downstream from the meter, the presence of valves and accessories which might distortion the velocity profile, etc.).

In the cases where the device is protected from suspended solids by filters, it will be recommended to register their existence in order to verify in the future their efficacy.

It is undeniable that this amount of information requires a considerable effort from the utility. However, the benefits on the short and medium term should compensate by far the investment in time and resources. Additionally, the largest part of the work is concentrated in an initial phase when most of the information is gathered and introduced in the database. After that initial period, it will only be necessary to update the data in the database when new users are connected, when meters are replaced or when conditions change.

Ideally speaking, a utility should be able, by means of a geographic information system of some sort, to relate the location of each meter with its characteristics stored in the database. With such a system it would be possible to establish the geographic relationship between variables such as the quality of water, the number of pipe breaks, the pressure in the distribution network and the rate of decay of the meters.

The final issue that the audit should address is whether the meter is the right size for the user. The circulating flowrate should ideally be inside the range defined by the transitional and the nominal flowrates, avoiding lower or higher values.

Domestic users do not usually present any difficulties in the selection of the right size of meter. However, medium and large meters that usually supply non-domestic users should be selected according to their capacity and nominal flowrate and not by the service pipe diameter. Any selection rule relating the size of the meter and the diameter of the service pipe should be avoided as a general principle. The only reliable method to determine if the capacity of a meter is the adequate one is to establish a flowrate pattern, either by means of a flow meter or an electronic meter with registering equipment. In case this information is not available, the peak flowrate can be estimated by means of sizing tables which relate it to the expected consumption volumes and the characteristics of the installation.

5.6.   MAINTENANCE AND RENEWAL OF METERS IN A UTILITY

5.6.1.   Maintenance of meters

Nowadays, and due to the low cost of water meters, the maintenance of metering devices is almost non-existent, and often limited to their periodic reading, cleaning of the filters and occasional replacement of certain elements (such as totalizers). For this same reason, the workshops where utilities used to repair and re-calibrate the meters have almost disappeared. Repair tasks as such are only economically justified these days in combined meters. The by-pass valve used in these meters (Chapter 2) needs to be completely watertight in order to guarantee a correct measurement. The valve consequently needs periodic cleaning. This is one of the reasons leading to a lesser use of combined meters which are being replaced by less demanding technologies.

In meters where there is an electronic component, its higher cost may lead to the mechanical part replacing only. Some companies manufacturing electronic meters offer contracts in which the replacement of the metering module and the battery are included after a certain number of years. The cost of this operation, including the necessary verification is quite lower than the price of a new meter. This procedure reduces considerably the yearly cost of these more sophisticated meters. If in the future the electronic devices maintain these higher prices, maintenance tasks may become more relevant again in order to reduce overall costs to the utility.

5.6.2.   Water meters replacement schemes

Replacing the meters in a utility in an orderly and rational manner is, no doubt, a complicated matter. Firstly, a clear distinction should be made between the abundant domestic meters and the meters that register the consumption on non-domestic users. The replacement of the first ones requires a probabilistic analysis, assuming that some of the replaced meters will be perfectly sound and that others should had been replaced long ago. Replacing a large number of meters in a utility requires a considerable amount of resources, and for this reason it is important to prioritize the replacement tasks.

The economic model presented in Chapter 7 helps to establish such priorities, and helps to determine the optimum renovation period for each model of meter depending on the characteristics of the user and the water supply. In any case, it is important to once again highlight the fact that depending on the consumption characteristics of the user, the optimum lifespan of a meter will be very different.

This is an advance reminder of the conclusions obtained in Chapter 7:

- The factors influencing the optimum renovation frequency are:
  - acquisition cost of the meter,
  - selling price of water,
  - discount rate,
  - rate of decay of the error curve,
  - water consumption, especially at low flowrates,
  - installation costs.
- A higher acquisition and installation cost of the instrument will increase the optimum replacement period.
- A higher cost of water or higher consumption rates will reduce the replacement period.
- The discount rate transforms future losses in current monetary units. Higher values will significantly increase the replacement period.
- A high rate of decay of the error curve will reduce the replacement period.

The combination of all these factors delivers a single figure for the renovation frequency of the meters for each meter model and type of user.

However, when dealing with non-domestic meters the approach to replacement must be completely different. Firstly, and due to the importance of every user out of the residential sector, a probabilistic approach is no longer recommended and every meter deserves an individual study and a unique renovation frequency. The data considered for these individual analyses should be collected directly from the meter and the connected users and never originate in statistical samples.

In any case, both the domestic and the non-domestic cases benefit from adequate audits of the metering system. Especially useful for this purpose is the commercial database that contains the billed consumption for each user. The statistical analysis of such information, along with the characteristics of users, installations and meters, allows identifying in many cases stopped or much deteriorated meters. From such information it is also possible to determine which models behave better or the influence of external variables. The quality of these analyses will greatly depend on the quality and quantity of the available data, and thus the convenience of accurate, up to date and comprehensive user databases.

A complete and periodic renovation of all meters in a utility requires careful allocation of resources, and consequently must be carefully planned in advance. Quite often utilities only replace meters which are suspected to be much deteriorated, and later perform massive renovation campaigns (many times with the cheapest available model). This procedure denotes lack of planning and can derive in important economic losses. Just imagine that the selected meter has a faulty design or is not adequate for the characteristics of the utility. The cheapest meter model may not be the optimum on the medium term either. As explained later in this book (Chapter 7), a higher initial cost of a more adequate model may be recovered in only a few years, especially taking into account the small differences found nowadays between meter models.

It must be emphasized that in order to carry out a massive replacement of meters with a single model, it is absolutely necessary to determine the behaviour of such model working with the specific characteristics of the utility. Some meter models have been known to fail consistently only under certain conditions (for instance the quality of water) just 2 or 3 years after they were installed.

For this reason, when planning the renovation of meters it is recommended not to limit the choice to only one model. The diversification is recommendable, and similarly to any other investment, it reduces the risks while providing valuable information on which meter model adapts better to the specific conditions of the distribution system.

A further recommendation, which will enhance future studies, is that the different meter models are randomly installed to users. Proceeding in such a way will enable reliable statistical studies to determine the error curve and compare the behaviour of models which have been simultaneously installed in the utility.

5.7. CONCLUSIONS

The management of water meters comprises several aspects which influence the overall performance of the metering system in a utility. Meters should be selected according to their working conditions. Whether it is a general decision of a domestic meter model, or a specific meter for a large industrial or commercial user, the selection criteria should be carefully studied and the demand characterized as detailed as possible.

In those cases where there is a great uncertainty about the optimum decision (for instance when making a selection for several thousand users), it is convenient to field test the meters and diversify the options in order to determine the best option and maintain it in the future.

Once the meter has been selected, the utility should make sure that the actual devices comply with the manufacturing specifications. Additionally, the appropriate steps should be taken to ensure that the installation of such meters is performed in such a way that their metrological characteristics are maintained.

Once the meter is in place, management procedures should ensure that the evolution of the meters' performance is known, and the optimum renovation frequency is determined.

6

How accurate are your meters?

6.1. INTRODUCTION

As shown in Chapters 2 and 3, the accuracy of water and flow meters of all types depends on the actual flowrate circulating through them. In other words, the error of any water meter is not a constant, and it changes throughout the measuring range (Figure 6.1). For medium and high flowrates the variations are minimum, but for low flowrates the error curve is steep until reaching the minimum flowrate.

Consequently, it is impossible to determine the percentage of water that a meter will register correctly unless the circulating flowrates are also known. And for the same meter, depending on the characteristics of the end user, the consumption could be registered with a great accuracy or with a significant error. For instance, the meter corresponding to Figure 6.1 (with a starting flowrate of 30 l/h) when installed in an empty household with a toilet leak of 20 l/h will not register any consumption (total volume per day will be 480 l). This same meter in a household with no leaks, and consumption flowrates ranging from 300 to 3000 l/h would measure the consumed volume with an error better than ±2%.

Several studies, including some by the authors (Arregui et al., 2003), show that the wear of mobile components in the meter has a greater effect in the accuracy at low flowrates. Usually, this sort of consumption is originated in leaks and toilets inside the households or in intermediate storage tanks.

Taking into account these facts, we can state that in order to assess the overall accuracy of all meters in a utility, it is necessary to have two sets of information. Firstly, the

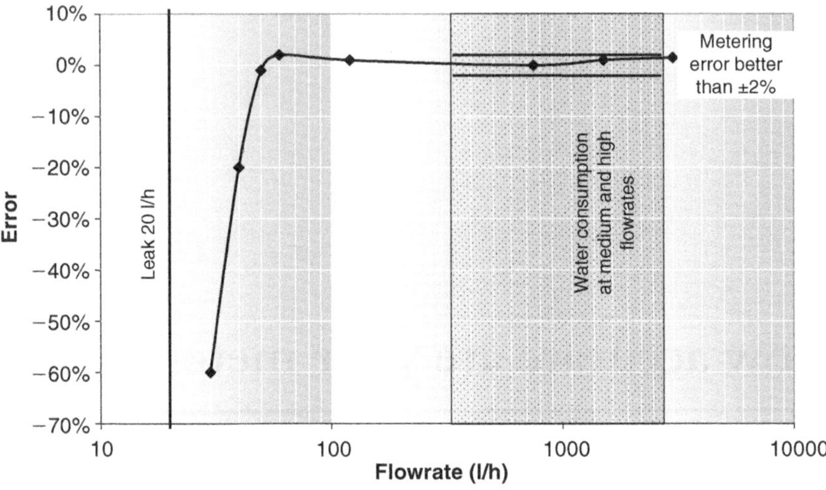

FIGURE 6.1.   ERROR CURVE OF A METER

error curves of all types of installed meters, and secondly, the demand flowrates of all different types of users.

The first set is easier to obtain, since it "only" requires a laboratory with a test bench. However, determining demand patterns is a more costly task and it is subject to greater uncertainties. The AWWA Research Foundation issued a report in 1993 (Bowen et al., 1993) that included average values for American users (and it can be used as a starting point everywhere else in the world). Since then, and in part due to the technological advances, several articles have appeared detailing the methodology to be followed and some reference values. However, in all cases, the extent of the case studies is quite limited, making impossible to extrapolate conclusions to other utilities. This lack of real and reliable data is, as a matter of fact, the greatest source of uncertainty when calculating the real accuracy of meters.

The following steps may be taken in order to determine the fraction of water not measured by the meters. First, meters must be clustered according to their characteristics (model, technology, age, nominal diameter, nominal flowrate, accumulated volume, etc.). A distinction between domestic and non-domestic meters must also be made. While the first should be studied from a statistical point of view, the second must be analysed individually, especially for large diameters.

Secondly, the users of the system must be classified according to their expected demand, taking particularly into account those consumption parameters in the lower and upper ranges of flowrates. The classification must be done considering parameters such as the state of conservation and the quality of the piping, the presence of intermediate tanks (Figure 6.4) and the size and type of household or facility. However, determining the demand pattern in a precise form for all these variables is extremely costly, and it is usually advisable to reduce the number of variables and focus on those with a greater impact: domestic tanks and the typology of the household (whether it is an apartment or a house with outdoor water use). As before, water consumption patterns of large consumers must be studied individually, due to the great heterogeneity of this class.

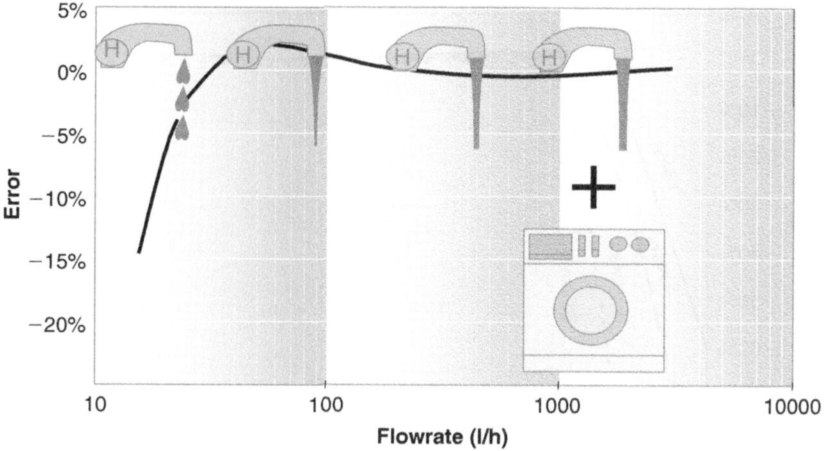

FIGURE 6.2. FLOWRATES AND METER ERROR DEPENDING ON THE END USE

Once these groups have been generated, for both meters and users, a data sampling should be carried out. It is important to point out that the conclusions obtained from the samples extracted from these groups will be extrapolated to the whole utility. Consequently, this process turns out to be critical for the final uncertainty of the study. Later, the results obtained from the water consumption samples must be crossed with the estimated meter error at different flowrates for each class of meter, in order to obtain the weighted error of a meter class installed in a given type of user. The uncertainty associated with each value of weighted error will be conditioned by several factors, such as the size of the samples and the heterogeneity of the groups (meters and users).

6.2. DETERMINING THE DEMAND PATTERNS

Consumption flowrates are extremely important in determining the weighted error of a meter. This parameter, which in practice is difficult to determine with accuracy for domestic users, becomes even more important for medium and large users in order to decide on the optimum size of the meter. The problem originates in the fact that for large users the demand patterns are more heterogeneous and should be obtained individually. For this reason, this section will mainly focus on the determination of the domestic demand patterns which are more homogeneous. However, the methodology to be used to determine the water consumption pattern of large users is exactly the same, without the need of statistical sampling.

The water demand in residential areas is well defined, and has its origin in very specific uses: faucets, washing machines, dishwashers, showers, toilets, leaks and outdoor uses. Figure 6.2 shows the flowrate range where a meter is working as a function of the type of end use. From the accuracy curve it can be clearly seen that the consumptions that pose a greater difficulty for the meter are leaks, normally associated to low and continuous flowrates.

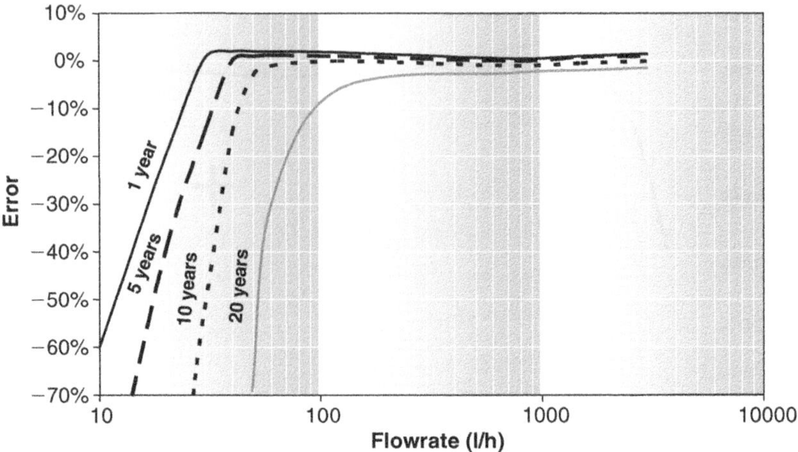

FIGURE 6.3. TYPICAL DECAY OF THE ERROR CURVE OF A WATER METER

Additionally, it is important to take into account that the error curve is generally stable in time at medium and high flowrates. It is, however, in the lower flowrates, where the largest errors appear, and the meter suffers more from wear (Figure 6.3).

These previous considerations show that in order to forecast the accuracy of a meter, it is vital to determine the volumes consumed at the lower end of the meter's range of measurement, for they will have the greater influence in the weighted error of the meter. Any variable affecting such volume will have a direct incidence on the weighted accuracy of the meters in a utility.

6.2.1. Past experiences in demand pattern determination

Several studies have been published to date on the determination of domestic consumption patterns. One of the most significant ones (Bowen et al., 1993) was published by the AWWA Research Foundation. The project carried out continuous measurements of consumption in 706 households located in five US cities.

The results of the project, however, were quite different to the ones obtained in previous studies. Especially those concerning the percentage of volume assigned to the lower range of flowrates (probably due to the metering equipment used and the number of households studied).

From the early 1960s to the late 1980s, several projects were started to register actual residential consumption (Table 6.1). These studies were even undertaken at a great scale (2300 households were studied by the AWWA in 1963) but the registering equipment was mechanical. The first work to use electronic data-loggers was carried out by Yanov and Koch in 1987, but only 12 households were monitored, a clearly insufficient number. The project carried out by Bowen used both electronic data-loggers and volumetric meters, allowing for a greater sensitivity at low flowrates.

The authors (Arregui, 2002) have also carried out characterization projects in Spain similar to the ones described in Table 6.1. Measurements were made in different types of households (with and without residential tanks) in order to identify the consumption

TABLE 6.1. DEMAND PATTERNS (PERCENTAGE OF WATER USE VS. FLOWRATE)

Reference	Year	Flowrate (lpm)							
		<0.95	<1.9	<3.8	<7.6	<15.1	<22.7	<37.8	>37.8
		Flowrate (gpm)							
		<0.25	<0.5	<1.0	<2.0	<4.0	<6.0	<10.0	>10.0
Kuranz	1942	13.6%	1.8%	5.0%	11.8%	52.4%	14.7%	0.7%	0%
Graeser	1958	5.0%	6.0%	8.0%	31.0%	40.0%	10.0%	10.0%	0%
Hudson	1964	13.0%	3.4%	6.8%	13.3%	43.0%	20.5%	20.5%	0%
AWWA	1966	4.6%	5.9%	5.9%	13.7%	59.0%	59.0%	16.8%	0%
Sisco	1969	1.0%	4.0%	4.0%	81.0%	81.0%	4.0%	10.0%	0%
Nielsen	1969	8.0%	8.0%	11.0%	18.0%	39.0%	20.0%	20.0%	4.0%
Brittain	1970	2.6%	1.8%	10.0%	21.9%	33.5%	19.7%	10.5%	0%
Yanov	1987	2.6%	2.3%	1.2%	8.0%	28.7%	20.5%	18.9%	17.8%
Bowen et al.	1993	8.8%	1.2%	3.1%	27.6%	14.8%	28%	13.1%	3.4%

TABLE 6.2. DEMAND PATTERNS IN SPAIN BY TYPE OF HOUSEHOLD

Flowrate	Type I (%)	Type II (%)	Type III (%)
0–12 l/h	4.7	10.0	2.7
12–24 l/h	2.8	3.1	1.9
24–36 l/h	1.9	1.8	1.6
36–72 l/h	4.3	4.2	4.5
72–180 l/h	8.5	11.6	5.7
180–1500 l/h	75.7	69.3	63.6
1500–3000 l/h	1.9	0.0	17.3
>3000 l/h	0.2	0.0	2.7
Average consumption	Approximately: 500 l/day	Approximately: 500 l/day	Approximately: 1200 l/day

behaviour of the users. These measurements were classified as follows depending on the characteristics of the household:

- *Household Type I*: Apartment blocks with direct injection from the network or a pump; 389 households were monitored for a week.
- *Household Type II*: Apartment blocks fed from an elevated tank (at the top of the building); 58 households were monitored for a week.
- *Household Type III*: Independent houses with garden. The summer consumption was monitored for over 4 weeks in 34 households.

The figures obtained by Bowen and Arregui (Type III) are quite similar for equal types of households.

Table 6.2 also presents some interesting (and apparently not logical) data. Type III households (larger and with garden) present a lower percentage of consumption at low

FIGURE 6.4.   DOMESTIC TANKS ON THE ROOFS OF BUILDINGS IN SPAIN

flowrates. These consumptions are related to leaks inside and outside the buildings after the meter, and it would be logical to assume that the larger the house, the greater the number of leaks. And that is exactly what happens. However, the overall consumption is greater, deeming a lower percentage of leaks (although in absolute terms, they may be greater).

Regarding higher consumption flowrates, it may be stated that, except in the case of houses with gardens, the demands very rarely exceed 1500 l/h, which is the typical nominal flowrate of domestic meters.

6.2.2.   Parameters affecting the demand patterns

The flowrates demanded by users, especially those in the lower range, establish the accuracy in the measure. Any foreign element or event that changes that flowrate will affect the capability of the meter to accurately register water consumption.

Amongst the elements affecting directly the volume of water consumed at low flowrates, two of them stand out: private domestic tanks (Figure 6.4) and leaks inside the households (usually in faucets and toilets).

The influence of domestic tanks in the quality of the measurements is very high, reducing considerably the ability of the meters to correctly register the consumption. Quite often a flapper valve controls the entry of water into the tank, and, as a consequence, only large demands, with a considerable drop in the water level, will force the valve to open completely. A partially open valve will let run into the tank water at low flowrates and will force the meter to work in the lower part of the measuring range. The solution to this problem lies in changing the flapper valve for one that can only be fully open or fully closed.

Toilet tanks are another source for low flowrates. The valves placed in these devices operate in the same way as the flapper valves described for the private domestic tanks. Usually, toilets can account for 1–2% of the demand at low flowrates, although nowadays it is quite frequent to see toilets equipped with instantaneous closure valves (toilet 2, Figure 6.5).

Low flowrates can originate in virtually any leak inside the household. It must be taken into account that users will only repair leaks when they are visible or noticeable, and higher-flowrate leaks are less frequent.

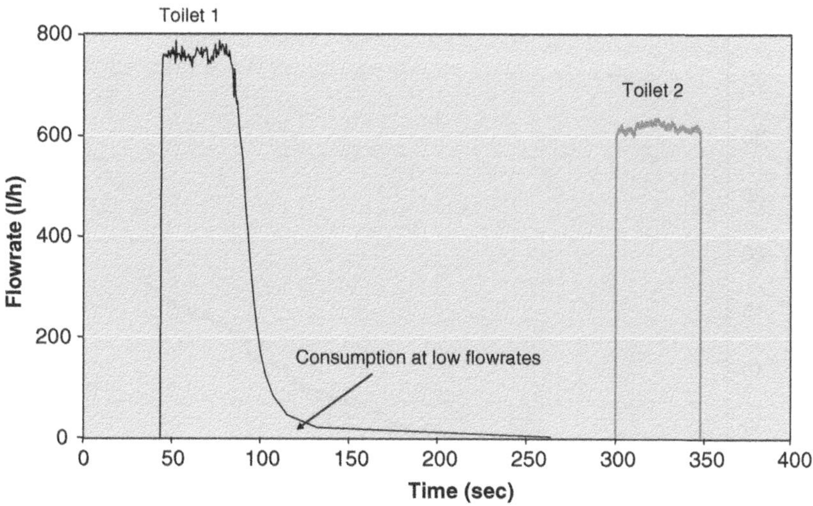

FIGURE 6.5. FLOWRATE PATTERNS OF TWO TOILET TANKS WHILE FILLING UP DEPENDING ON THE INLET VALVE CHARACTERISTICS

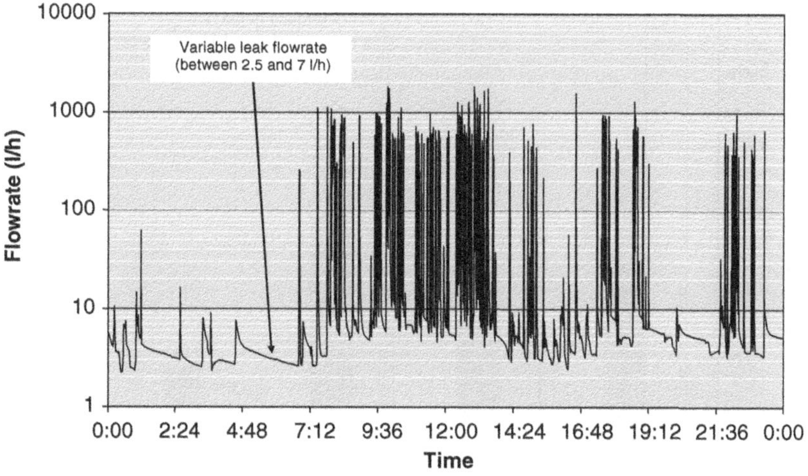

FIGURE 6.6. DEMAND PATTERN FOR A HOUSEHOLD WITH A LEAK

Often, leak flowrates are a function of the network pressure and usually do not remain constant through the day (Figure 6.6). The meter may then register leaked volumes at times during the day (night hours), and not be able to measure them at others.

Leaks in households can be quite small, although larger leaks are also found occasionally. The authors found out during field measurements (Arregui, 2000) that most of them ranged from 2 l/h (the minimum flowrate of the meter) to 40 l/h. Exceptionally, some of them went all the way up to 100 l/h. Figure 6.7 shows the

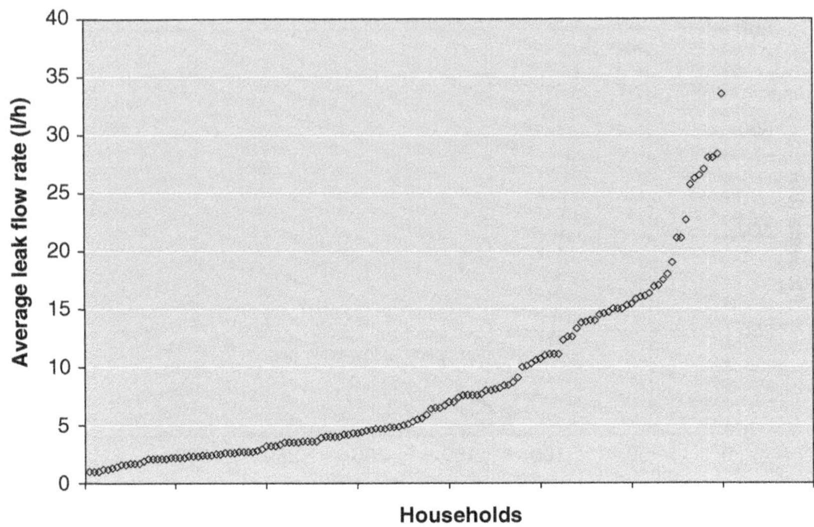

FIGURE 6.7.   AVERAGE LEAK FLOWRATE PER HOUSEHOLD

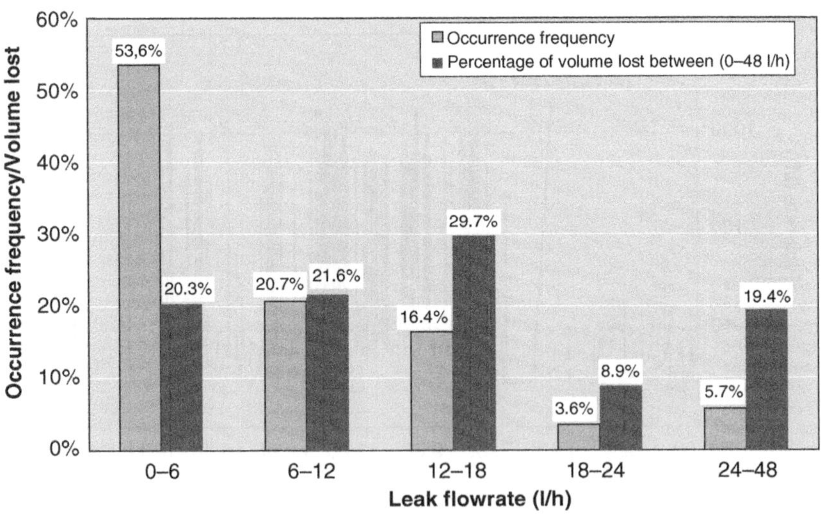

FIGURE 6.8.   LEAK FREQUENCY IN HOUSEHOLDS AND TOTAL LEAKED VOLUME

individual average leak flowrate of each household, while Figure 6.8 shows the statistical distribution of these leaks.

Another important factor in a user's demand pattern is the volume consumed at flowrates at the top of the meter's range. If the meter is operating frequently at these flowrates, the instrument can deteriorate faster. In these cases, the meter should be replaced by a higher-capacity instrument. For domestic users, this circumstance only takes place in residential properties with gardens, as shown in Table 6.2. In apartment blocks, the volume consumed at flowrates above 1500 l/h is almost negligible. This same conclusion was reached by the study carried out by Bowen, in which these types of households held 95% of their consumption at flowrates below 2000 l/h.

Finally, it is necessary to mention other variables, such as the temperature, rainfall and the seasonal behaviour of the users, which affect their demand patterns. However, due to the difficulties implied in taking into account these variables, no specific conclusions have been reached about the actual influence on the domestic demand patterns.

6.2.3. Obtaining demand patterns at the field

Although on a first approach the demand patterns shown previously can be used, to obtain reliable results it is recommended that the actual demand patterns for the users of the utility are obtained. For instance, the quality of water or frequent supply interruptions may increase leaks in households and as a consequence increase the volumes consumed at low flowrates. Especially, commercial and industrial patterns (much more heterogeneous in nature) should be handled individually.

The first step in determining the domestic consumption patterns is to classify users by type. The classification should be made according to simple criteria so the households are easily assignable to one category. Usually, two or three categories will be enough to obtain good results without increasing excessively the costs of the project.

Since the demand patterns are obtained from a sample of users, the quality of the patterns obtained is directly linked to how representative the sample is. The households chosen for the measurements should be characteristic of the category of users they belong to. Should the consumption of these households deviate from the average consumption of the category, they would affect the reliability of the study.

6.2.4. Monitoring equipment

The actual monitoring of the consumption should not distort the demand itself. The equipment needed to do this is basically an accurate meter equipped with a pulse emitter, and a data-logger (although other combinations are possible) (Figure 6.9).

The meter used in this sort of setup should be, at least, a Class C volumetric meter with a very low starting flow, which is able to register leaks. Then, ideally, the distribution of consumptions should be corrected with the error curve of the meter. This correction may turn out to be quite important at low flowrates, where the percentage of registered consumption may be affected significantly by the error of the meter.

As the consumed flowrates are registered, they need to be transmitted to the data-logger. This process may also be an important source of distortions. If pulse emitters are used, the associated volume to each pulse should be as low as possible, for instance 0.1 l or less. Additionally, the pulse emitter should not affect the error curve of the meter (some Reed pulse emitters are known to increase the starting flow of the meter due to the need of placing a magnet in one of the gears of the totalizer, and thus increasing friction).

If the volume associated to each pulse is too large, the registered volume at low flowrates would increase. This is due to the fact that the flowrate associated to each pulse is calculated from the time lapsed between the previous pulse and this one. If the volume for each pulse is too large (for instance, 1 l), a single short demand would not be able to

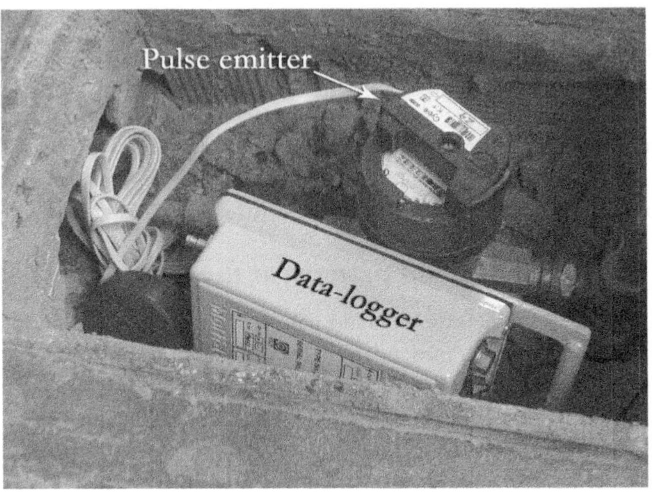

FIGURE **6.9.**   WATER DEMAND MONITORING EQUIPMENT

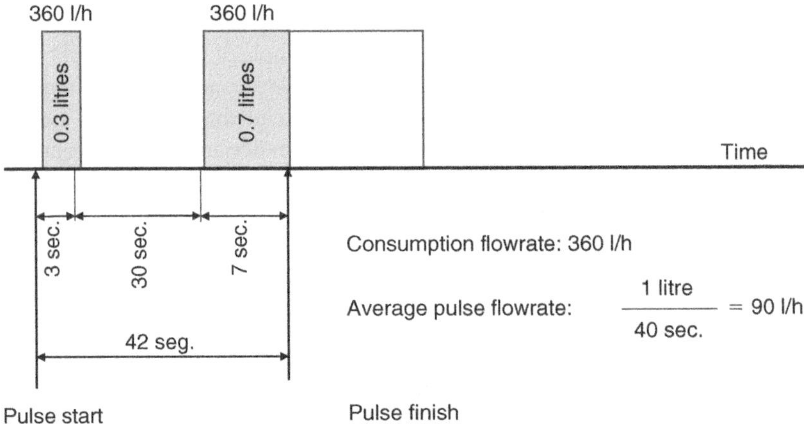

FIGURE **6.10.**   EXAMPLE OF THE DISTORTION IN THE DETERMINATION OF THE DEMAND
FLOWRATE WITH LOW-RESOLUTION PULSE EMITTERS

generate several pulses. The volume associated to a single pulse would then be distributed between several demands, lapsing a longer time than the one really used to create that demand (the same volume "apparently" takes longer time to be consumed). Figure 6.10 shows how two very short demands of 360 l/h can be accounted for as a single demand of 90 l/h. Details of how this parameter affects demand patterns can be found in Arregui (1999).

Inductive pulse emitters have proven to be the most reliable ones in the tests carried out by the authors. The advantages they provide are:

• High commutation frequency (above 10 Hz). The resolution of the pulse emitter can be improved, obtaining volumes of less than 0.1 l per pulse. Reed emitters on the other hand can only work at a maximum frequency of 1 Hz, limiting for instance the minimum volume per pulse in a 15 mm to 1 l.

- Most commercially available models allow to determine the direction of the flow. This fact is important in the absence of a non-return valve, with the possibility of water flowing back to the network.
- Unlike Reed emitters, which are mechanical, inductive emitters do not create false signals when the switch bounces.
- Less sensitive to electromagnetic interferences.
- Usually quite robust, reaching very often protection degrees of IP68.
- They are independent and do not need external power supply (batteries).
- They do not affect the sensitivity of the meter at low flowrates.

Data-loggers have evolved in the past few years, and the memory limitations of the past have now been overcome. In any case, the memory requirements of a logger for the uses explained in this book are not very large, a model with 1 MB of memory will allow to register over a month worth of data, of a reasonable resolution, from four users.

Loggers usually store the number of pulses emitted in every time interval (i.e. 10 sec). However, they can also store the occurrence of every pulse in time (this method will obviously consume more memory but reduce distortions in the demand pattern).

The main factors to take into account when buying a data-logger for these applications are:

- Total memory of the logger. Memory storage modes.
- Availability of analogue inputs (4–20 mA, 0–5 V, etc.).
- Number of input channels.
- Robustness.
- Battery life.
- Size (important for installation in tight spaces).
- Theft and manipulation protection.

With the latest developments in technology, different alternatives to data-loggers have started to appear. For instance, PDAs could be used to perform similar tasks at a much lower price and much larger capacity and computing capabilities. However, these devices still present problems in terms of battery life and robustness, although they have clearly become an alternative.

6.2.5. Other types of equipment

There are other possibilities when it comes to obtaining the demand patterns of users. They are generally related to the use of meters with electronic totalizers, since most commercial models allow users obtain the value of the flowrates that have circulated through them.

This method however poses several inconveniences. For instance, the intervals in which the flowrate is divided are not definable by the user, and they are often too few in the lower range. Additionally, this type of totalizers is not used in volumetric meters and can be found in paddle wheel, single jet and multiple jet meters, with starting flowrates too high for the task (leaks would be left out). However, this disadvantage is less of a problem nowadays, when some manufacturers have introduced single jet meters with *theoretical* starting flowrates of 3 l/h. Finally, the information provided by the meter could be described as a "summary" of the flowrates that circulated the pipe. The meter does not store data continuously and only provides statistics on water demand.

This option allows to obtain consumption patterns from a large number of households at a much lower price. However, the quality of the information will be much lower (and the advantage of quantity vs. quality is certainly not clear in this case) than the one obtained with the equipment explained above.

In any case, the main application of this type of device is to determine water consumption patterns of large users.

6.2.6.   Duration of the monitoring

A key issue in this process is the amount of time that a user needs to be monitored. In practical terms, this parameter is set by the budget of the project and the amount of available registering devices.

As a general rule, the demand of domestic users in the absence of leak remains constant in time, and a week's monitoring should be more than sufficient. However, when leaks are present, the consumption flowrates change significantly depending on the total volume consumed. In these cases, the recommendation would be to maintain the monitoring for 1 or more weeks. In practice, not knowing beforehand if a user will or will not have leaks in the household, the recommended time is at least 1 week for all users.

6.2.7.   Conclusions

- Consumption flowrates modify the percentage of water accounted for by the meter. The same instrument may behave, from a metrological point of view, differently in two users with different demand patterns.
- Residential demand patterns should be obtained from samples at the users' households. An adequate classification of the types of users, and choosing representative households for the sample, is essential.
- Monitoring equipment should be carefully chosen to avoid distortions of the measured consumption patterns with respect the actual demand pattern. The use of volumetric meters and high-resolution pulse emitters (0.1 l per pulse or less) is recommended.
- In the case of choosing meters with electronic totalizers, that provide information about demand patterns, the metrological quality of the devices should be carefully inspected. The information that will be obtained in such cases will be more specific and easier to handle, but poorer in quality and less exhaustive.

6.3.   OBTAINING THE ERROR CURVE

The other parameter that can affect the percentage of registered volume besides the consumption pattern is the error curve. Its shape defines the capacity of the instrument to measure water volumes throughout its life. Not all metering technologies (not even all meters with the same technology) have the same rate of decay of the accuracy with time. Additionally, external parameters such as the quality of water may influence each model in a different way. The error curve, and the way it evolves, must consequently be obtained for every meter type specifically.

However, obtaining the error curve is not as easy as obtaining the consumption pattern of a user. The number of tests needed to obtain a well-defined curve is too large and the process too costly. The procedure followed, in practice, implies testing only a few selected flowrates and reconstructing the rest of the curve from that information. The selection of these flowrates is often overlooked despite its importance, as it will be shown later.

Once the test flowrates have been selected, the sampling must be planned in order to limit the uncertainty to a desired level while covering the different models and ages groups of the tested meters. It is important to always bear in mind that the results obtained for the error curve will always originate from a sample of meters, and the conclusions are always estimates with associated uncertainties that should not be forgotten.

One apparently minor, and yet important, note must be added. From the moment the meters are collected from the network until they reach the laboratory, they should always be kept full of water to avoid them drying. Otherwise the results obtained in the laboratory could be distorted. This and other important details will be covered in the chapter devoted to meter testing.

6.3.1. Selection of flowrates for the tests

The impossibility of testing meters for a large number of different flowrates creates the need to reconstruct the error curve from the error figures at a few different flowrates (AWWA Manual M6). Traditionally the meter is tested at three different flowrates. One at the lower end of the range, one in the middle and one at the higher end of the range.

Domestic meters

Several authors, like Yee (1999) and Allender (1996), use the AWWA standards recommended testing flowrates, that is ¼, 2 and 15 gpm. These standards cover meters with flowrates ranging between ¼ and 20 gpm. These flowrates have been chosen so the error at ¼ gpm represents the errors found between 0 and ½ gpm. Likewise, the error at 2 gpm covers the range between ½ and 2 gpm and, finally, the error at 15 gpm accounts for any flowrate above 2 gpm. Figure 6.11 shows the reconstruction of the error curve from these values. It is easy to notice the great differences between the reconstructed curve and the real one. This implies that the ¼ gpm flowrate is not representative of the errors of this meter at the lower end of its range.

Sometimes, this loose match between the reconstructed and real curves is justified saying that the consumption at the lower end of the range is very small (Yee, 1999). However, this statement is not always correct (Tables 6.1 and 6.2). Furthermore, the older the meter becomes, the wider the range of misrepresented flowrates gets, making it more difficult to correctly estimate the rate of decay of the instruments.

Experience has shown that it is at the lower flowrates where the measuring errors are more important, and similarly to the consumption patterns, the error curve must

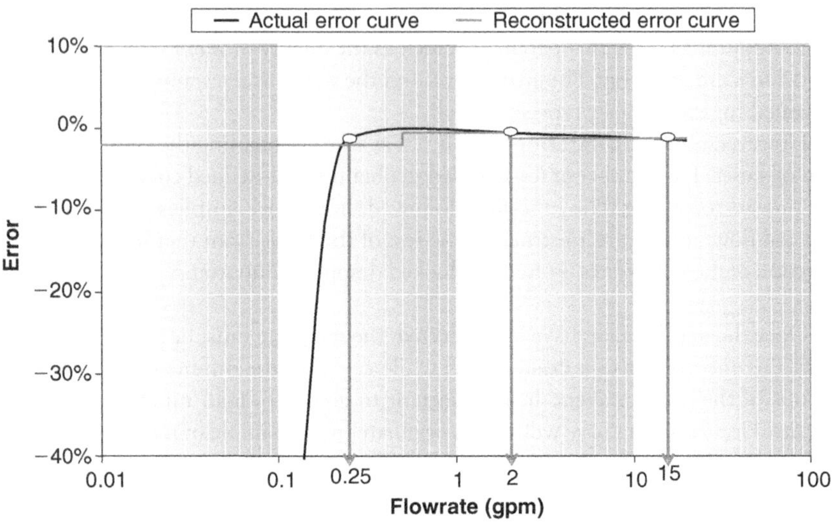

FIGURE 6.11. RECONSTRUCTED ERROR CURVE USING AWWA'S SUGGESTED FLOWRATES

be very well defined for these flowrates. In order to correctly reconstruct the error curve in this range it is important to know or closely estimate the starting flow. It is also recommended to test the meter at the minimum flowrate (ISO 4064) and at another flowrate between the minimum and the transitional values. Finally and taking into account that the consumption flowrates rarely go above 1500 l/h, and that the error at medium and high flowrates is almost the same, tests at 750 and 1500 l/h would be enough. It can be considered that the error from that flowrate onwards remains reasonably constant up to the maximum flow. The errors at all other flowrates may be obtained by a linear interpolation between the available points. Figure 6.12 shows the reconstruction of the error curve of a meter using the described method, which reproduces the real curve much more accurately than the previous procedure.

This technique, however, presents several disadvantages. The first one is that there is no universal definition of "starting flowrate". It should be noticed that the flowrate needed to initiate the motion of the impeller is different from the flowrate that is needed to keep it in motion, and both could be considered "starting" flowrates. Frequently, this last figure is the one that is useful, for it is the threshold where the meters continue registering leaks after consumptions at medium or high flows.

The second disadvantage is that quite a long time is necessary to determine the starting flowrate in a laboratory. This figure can be estimated from the errors found at low flowrates. Figure 6.13 shows the relationship between the starting flowrate and the error at 30 l/h in single jet, Class B meter. This type of correlation allows estimating the starting flowrate and reconstructing the error curve from the 30 l/h error.

From all these facts, and for domestic users, the following test flowrates recommendations could be issued (Table 6.3).

Consumption flows of non-domestic users are difficult to predict and do not necessarily remain constant in time. Consequently, it is necessary to obtain the consumption

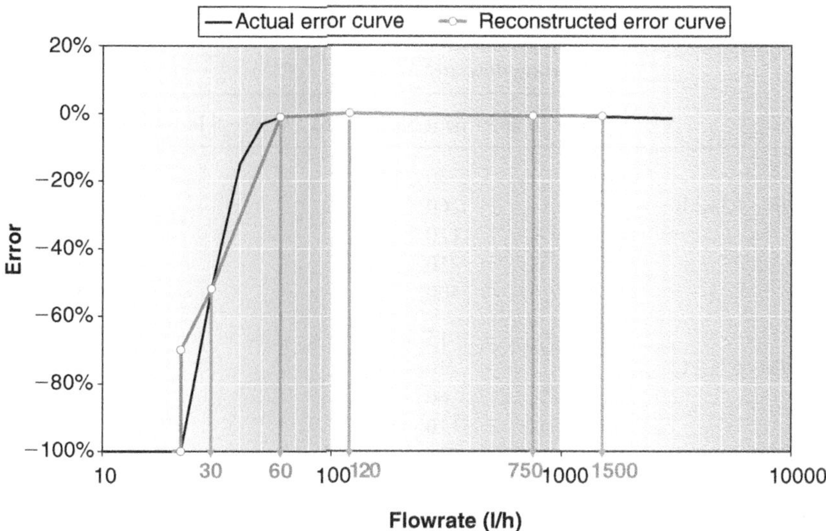

FIGURE 6.12. RECONSTRUCTED ERROR CURVE USING THE STARTING FLOWRATE AND LINEAR INTERPOLATION

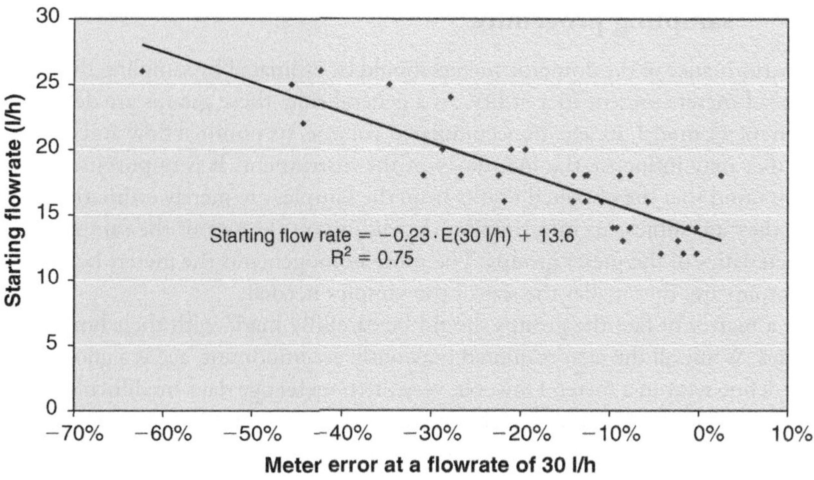

FIGURE 6.13. STARTING FLOWRATE AS A FUNCTION OF THE METER'S ERROR AT 30 L/H

pattern of the user in order to determine at which flows the meter will be operating. Once this has been achieved, the test flowrates should be distributed along the usual operating range of the meter.

Whenever the flowrates circulating through the meter cannot be obtained, the test values should be the minimum, the transitional, the nominal and the maximum flowrates, covering all the actual measuring range of the instrument.

TABLE 6.3. Recommended test flowrates for domestic users

	Starting flowrate*	
	Flat in apartment blocks (l/h)	Houses with garden (l/h)
Meter	30	30
1.5 m³/h – Class B	60	60
	120	120
	750	1500
	1500	3000
Meter	15	15
1.5 m³/h – Class C	30	30
	60	60
	120	120
	750	1500
	1500	3000

*Or an estimation from low-flowrate error values.

6.3.2. Grouping the meters: Setting up the sampling procedure

The performance of the domestic meters should be estimated by sampling the different groups of meters present in a utility. As a general rule, these groups are defined by a certain meter model, its age, its accumulated volume, its nominal flow and other variables that may influence the metrology of the instrument. It is important to always bear in mind that the obtained results from the samples are merely estimations and as such, they are subject to uncertainties depending on the size of the sample and the characteristics of the meter groups. The more homogeneous the meters belonging to each group are, the smaller the size of the samples needed.

As a matter of fact the groups should be carefully made with their homogeneity in mind. While all the factors quoted previously are important, age is a good parameter to define wear in a meter. However, very often meter age data are difficult to obtain or they are not very reliable. On the other hand, the accumulated volume is usually an up-to-date parameter which is carefully controlled by the utility since it is used for billing purposes.

When meters are classified by age, the groups can span 2 or 3 years. If the groups depend on the accumulated volume, they should be separated every 300–500 m³ (which would be equivalent to a 2–3 years average residential consumption depending on the user). In both cases the authors found an important correlation between the wear at different flowrates and age.

Once the meters have been grouped, random samples from each group are selected. Key variables such as the installation characteristics, the location, the characteristics of the consumer, etc. should be well represented in the samples and limiting the sampled meters to instruments with similar characteristics should be avoided.

TABLE 6.4. RELIABILITY FACTOR AT
DIFFERENT CONFIDENCE LEVELS

Confidence level (%)	$z_{1-\frac{\alpha}{2}}$
99	2.57
95	1.96
90	1.64
80	1.28

6.3.3. Analysing the results

For every tested meter, and for each flowrate used, an average error figure is registered. This average value, which is the mean value of all errors tested for a single flowrate, is an estimation of the real error for the whole meter group. The real value of the error for the whole group will belong to a range of values around that figure with a certain probability. Such interval is known as the confidence interval. Statistically, it is defined as the range of values where the real value of the estimated parameter is expected to be found with a certain probability (Table 6.4).

When the sample size (number of tested meters per model and age) is 30 or more, the confidence interval can be calculated as follows:

$$\varepsilon_{Min}^{Q} = \varepsilon_{Ave}^{Q} - z_{1-\frac{\alpha}{2}} \cdot \frac{s}{\sqrt{n}} \tag{6.1}$$

$$\varepsilon_{Max}^{Q} = \varepsilon_{Ave}^{Q} + z_{1-\frac{\alpha}{2}} \cdot \frac{s}{\sqrt{n}} \tag{6.2}$$

where:

s is the standard deviation of the errors at a flowrate Q.

n is the number of tested meters.

$z_{1-\frac{\alpha}{2}}$ is the reliability factor.

$1 - \dfrac{\alpha}{2}$ is the confidence level.

As deduced from the previous equations, the larger the samples are, the narrower the range and the smaller the uncertainties (Figure 6.14). The contrary happens with the standard deviation. When the group is heterogeneous (and "s" is large) the uncertainties will become greater. This explains why for the same sample size, the uncertainty is much higher at low flowrates than at medium and high flowrates (Figure 6.14). The dispersion of the results is much higher at low flowrates, for meters are then less accurate, and there is a large quantity of variables affecting metering performance.

Large diameter meters must be tested individually, and the sampling concepts explained previously cannot be applied. In these cases, the error to be found does not have a greater uncertainty than the instrument's own calibration and repeatability.

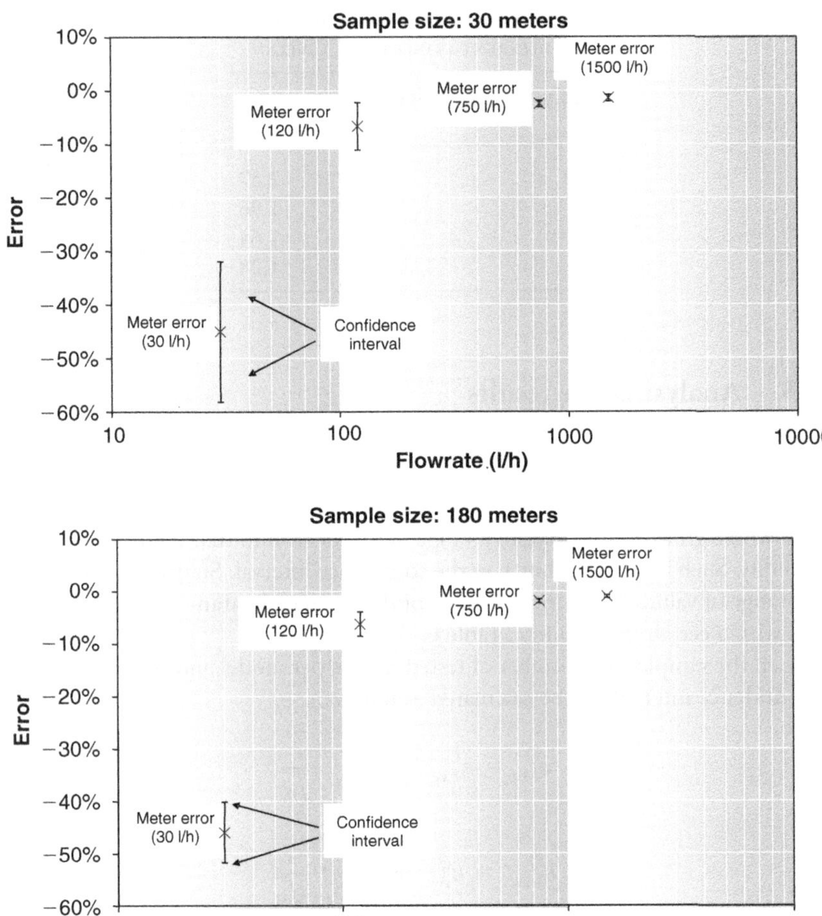

FIGURE 6.14. CONFIDENCE INTERVALS OBTAINED FOR A METER 10 YEARS OLD

6.4. CALCULATING HOW ACCURATE YOUR METERS ARE

Once the users' consumption patterns are known and the error curve of the meters has been reconstructed, it is possible to determine the percentage of the actual water consumption that these meters are registering. However, obtaining reliable results depends on the correct weighting of both parameters.

Some of the simplifications when combining these two parameters that are proposed in the technical literature, for instance the recommendations by the AWWA, may lead to inaccurate results as shown by Arregui et al. (1998). Generally speaking, these inaccuracies come from ignoring the starting flowrate, or more precisely speaking, the non-metered volume at flowrates below the starting value.

More specifically, the works published by Yee (1999) and Allender (1996) use the AWWA recommended flowrates to calculate the weighted error, that is ¼ gpm, 2 gpm and 15 gpm. The meters used in the USA (5/8″) for residential purposes have a larger maximum flowrate than the equivalents, used all around the world, according to ISO 4064:1993. Following similar criteria (minimum, transitional and nominal flowrates) the test flowrates for ISO 4064:1993 meters would be 30, 120 and 1500 l/h.

According to this method, the weighting would consist in multiplying the fraction of the volume consumed at each flowrate range by the corresponding error. For domestic users with the characteristic consumption pattern of an apartment block (Table 6.2, household Type I) the equation would read as follows:

$$\text{Weighted error} = 0.094 \cdot E(30\,\text{l/h}) + 0.128 \cdot E(120\,\text{l/h}) + 0.778 \cdot E(1500\,\text{l/h})$$

Being 0.094, 0.128 and 0.778 the fraction of the volume consumed at each flowrate range (their sum is 1). Obviously, the weighted error depends on the distribution of the volumes consumed at each flowrate. For instance, the weighted error for the meter in Figure 6.12 for the three types of households would be:

Household Type I:

$$\text{Weighted error} = 0.094 \cdot (-52\%) + 0.128 \cdot (0\%) + 0.778 \cdot (-1\%) = -5.7\%$$

Household Type II:

$$\text{Weighted error} = 0.149 \cdot (-52\%) + 0.158 \cdot (0\%) + 0.693 \cdot (-1\%) = -8.4\%$$

Household Type III:

$$\text{Weighted error} = 0.062 \cdot (-52\%) + 0.102 \cdot (0\%) + 0.836 \cdot (-1\%) = -4.1\%$$

As expected, the error is greater in those households with a larger volume lost through leaks as opposed to households where consumption at low flowrates is not so relevant.

Nevertheless, and as mentioned before, assuming that the total volume consumed at low flowrates is registered with the error of the meter at 30 l/h creates significant errors. In order to overcome this problem an alternative weighting method is proposed, based on the following presumptions (Table 6.5).

Calculating the weighted error for the meter of Figure 6.12 and a Type I household, the figures shown in Table 6.6 are obtained.

The same procedure produces errors of −14.9% and −6.5% for household Types II and III. These results are quite different compared to those obtained by the procedure described by Yee and Allender. It is important to point out that in both procedures the same consumption pattern and error curve (although reconstructed differently) were used. The reason is that the first method (AWWA) does not take into account the fact that part of the consumption at low flowrates takes place below the starting flowrate and is not registered (Table 6.7).

6.4.1.   Combined metering error of installed meters in a utility

Once the weighted error for a group of meters has been obtained, it is possible to determine the amount of non-metered water in a utility. This procedure requires determining

TABLE 6.5.   ASSUMPTIONS IN WEIGHING THE CONSUMPTION PATTERN AND THE
ERROR CURVE

Error curve (Figure 6.15)	Consumption pattern
• Metering error at the starting flowrate is −70%. • The error curve is supposed to follow a linear evolution between the known error values.	• Within every interval of consumption flowrates, the consumption is distributed uniformly.

Weighted metering error
• Any volume consumed at a flowrate below the starting flowrate is not registered.
• For every interval in the consumption patterns, volumes are registered according to the error corresponding to the mean flowrate of the interval, obtained from reconstructing the error curve (Figure 6.15).

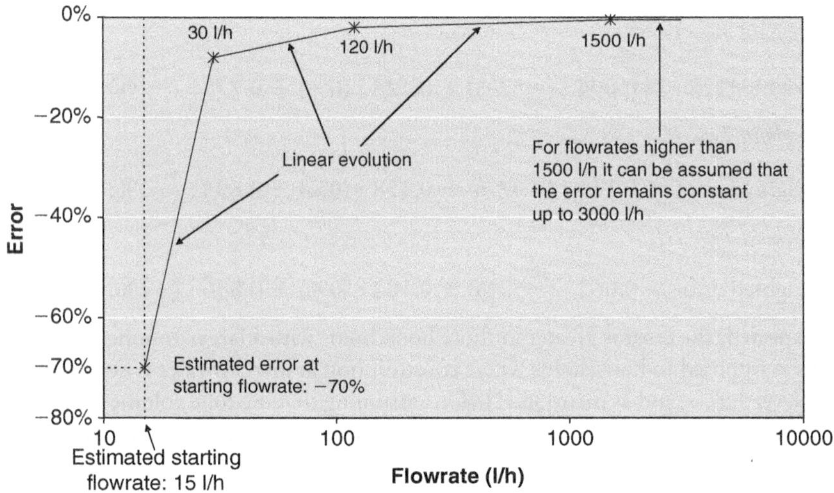

FIGURE 6.15.   RECONSTRUCTION OF THE ERROR CURVE FROM ERROR DATA AT
30, 120 AND 1500 L/H

the "weight" of each group in the total consumption of water. The weighting coefficients could be calculated in a first approach as:

• Percentage of meters in the group over the total number of meters.
• Percentage of volume registered by meters in the group over total registered volume.

However, the most accurate method consists of calculating the weighting coefficient for each group based on the percentage of the actual volume consumed by the users of that group with respect to the total volume consumed (adding together all the groups).

TABLE 6.6.   ASSESSMENT OF THE WEIGHTED ERROR OF A METER

Flowrates (l/h)	Household Type I (%)	Meter error (%)	Registered volume (%)
0–12	4.7	−100	0.00
12–22	2.3	−100	0.00
22–24	0.5	−68	0.16
24–36	1.9	−52	0.91
36–72	4.3	−11	3.83
72–180	8.5	0	8.50
180–1500	75.7	−0.8	75.09
1500–3000	1.9	−0.8	1.88
>3000	0.2	−0.8	0.20
		Total	90.6%
		Error	−9.4%

TABLE 6.7.   COMPARISON OF THE WEIGHTED ERROR OBTAINED USING BOTH WEIGHING METHODS

Household	Weighted error obtained using error values at three flowrates (%)	Weighted error taking into account the starting flowrate (%)
Type I	−5.7	−9.4
Type II	−8.4	−14.9
Type III	−4.1	−6.5

In order to obtain the actual consumption, the volume registered by each group of meters (i), corresponding to one category of meter and one type of user, is corrected with its weighted error (ε_i) expressed as per unit fraction:

$$\text{Consumed volume}_i = \frac{\text{Registered volume}_i}{1 - \varepsilon_i} \tag{6.3}$$

The total volume consumed by metered users will be:

$$\text{Total consumed volume} = \sum \text{Consumed volume}_i = \sum \frac{\text{Registered volume}_i}{1 - \varepsilon_i} \tag{6.4}$$

As a consequence, the weighted error for all the meters will be obtained as the ratio between the non-metered volume (or over registered) and the actual consumed volume:

$$\varepsilon_{\text{global}} = \frac{\sum \frac{\text{Registered volume}_i}{1-\varepsilon_i} - \sum \text{Registered volume}_i}{\sum \frac{\text{Registered volume}_i}{1-\varepsilon_i}} = 1 - \frac{\sum \text{Registered volume}_i}{\sum \frac{\text{Registered volume}_i}{1-\varepsilon_i}} \tag{6.5}$$

The following example shows how to calculate the metering error in a utility with 6200 users, and with two meter models of different ages:

Age	Error (%)	Number	Registered volume (m^3)	Consumed volume (m^3)
Model 1				
1–2 years	−6.5	900	129,295	138,284
3–4 years	−9	1500	204,445	224,665
5–6 years	−13	2000	281,479	323,539
			Consumed volume (m^3)	686,488
			Error Model 1	−10.4%
Model 2				
3–4 years	−5.7	200	24,158	25,619
5–6 years	−7.2	450	57,894	62,386
6–7 years	−8.9	650	86,874	95,361
8–9 years	−10.2	500	69,263	77,130
			Consumed volume (m^3)	260,496
			Error Model 2	−8.6%
			Total registered volume (m^3)	853,408
			Total consumed volume (m^3)	946,984
			Weighted meters error	−9.9%

The combined error of all Model 1 meters is −10.4%, although each age group behaves differently. The combined error of all Model 2 meters is lower than −8.6%. However, the consumption of users with a Model 1 meter is more than twice than the volume consumed by Model 2 users, and as a consequence the influence in the non-metered volume of Model 1 is greater.

6.4.2. Rate of decay of a meter's error

The evolution of the weighted error of a certain model is easily obtained when the meters are stratified in age groups. In general terms, a linear approximation reproduces the behaviour well enough, although in some cases it might not be adequate. Figure 6.16 shows the evolution of the error at four different flowrates of 220 velocity meters (nominal flowrate 1.5 m^3/h) as a function of the accumulated volume. The evolution could be described quite accurately by a linear function. However, in this case a second-grade polynomial fits the evolution even better.

Taking into account this error curve, simplifying the prediction with linear adjustments and for the consumption pattern of a household Type I, the weighted error follows the evolution shown in Figure 6.17.

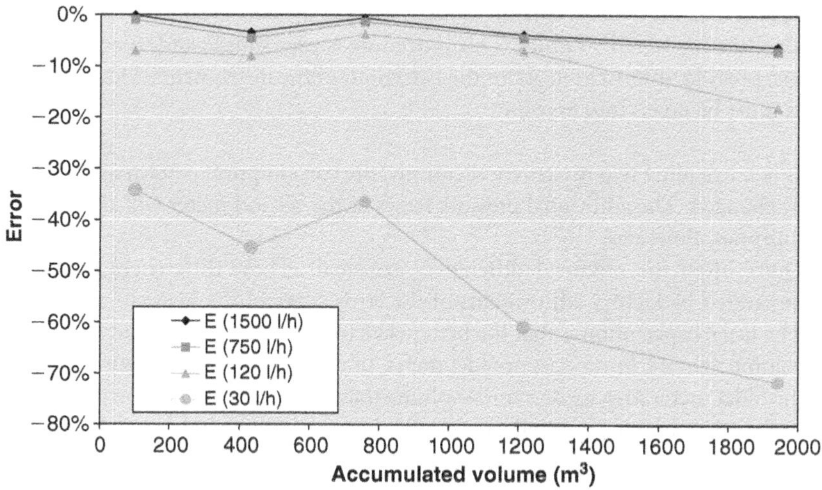

FIGURE 6.16. RATE OF DECAY OF THE ERROR WITH THE ACCUMULATED VOLUME AT DIFFERENT FLOWRATES

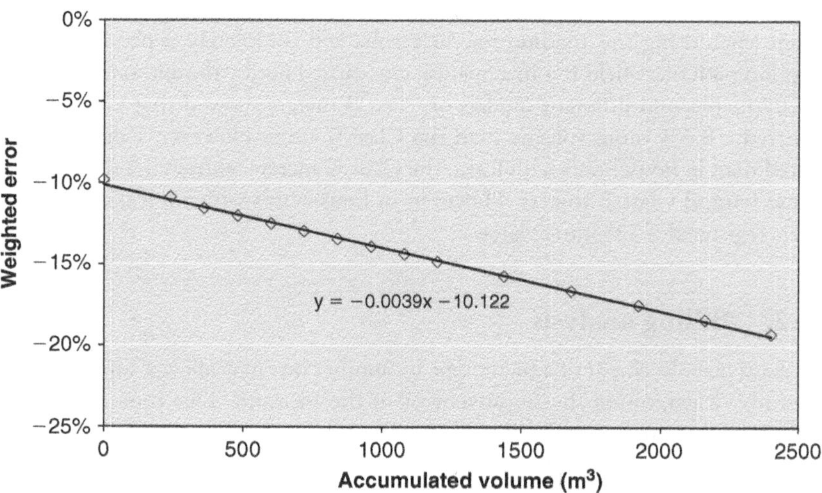

FIGURE 6.17. RATE OF DECAY OF THE WEIGHTED ERROR WITH THE ACCUMULATED VOLUME

6.5. OTHER METHODS TO ASSESS THE WEIGHTED ERROR OF A METER

6.5.1. Serial installation of meters

Quite often, the staff in charge of managing the meters of a utility use simpler methods to assess the percentage of volume that every model ceases to register (or to even compare two different meter models).

One of the most common methods is the serial installation of meters to later compare the registered volumes. This procedure is very simple and, used in real conditions, provides information about the behaviour of the instruments. However, several issues must be taken into account:

- When installing two meters consecutively, the consumption pattern of the user will be changed. The additional pressure losses of the second meter will affect the consumption flowrates.
- Quite often the obtained differences are small. These differences can even be motivated by factory adjustments of the error curve of the meters.
- The usual expectation is that the better class (or newer) meter should register more volume than the lower class or older meter. In practice, this is often not the case, and the older meter may register more volume than the newer one. The explanation can be found in the consumption patterns. Those users with no leaks in their households will have the meter always working at medium or high flowrates. At these flowrates, some meters (velocity meters) may register more volume than the real one (although at low flowrates, the tendency will always be to register less volume). As a consequence, the older meter in this situation will register more volume.

Taking into account these previous considerations, the user sample should be carefully chosen when using this method for the results will completely depend on the consumption pattern. A field test in a Spanish city carried out by the authors with Class C velocity meters and different models of Class B meters showed that Class B meters registered a 0.6% more volume than the Class C ones. However, a deeper analysis showed than in households with leaks, the Class C meters registered 3.6% more than the mechanical Class B meters. However, in households with no leaks, the Class B meters registered 2.7% more water.

6.5.2. Billing analysis

The massive replacement of a meter type by another one may allow a simple comparative study to determine the improvement in the measure. This procedure requires extreme caution, and typical differences are usually in the order of 5–6%.

As a general rule, the comparison period should cover the possible seasonal variations in water demand. It is in any case advisable to correct the bills from the new meter type with the evolution of the metered volume in users with the older model. This type of analysis will be covered in more detail in Chapter 7.

6.6. FACTORS AFFECTING THE QUALITY OF WATER CONSUMPTION MEASUREMENT

The following pages address the most significant factors that may influence the control of water consumption in domestic, commercial and industrial users. These factors obviously affect the selection of the most adequate meter and its renovation frequency.

6.6.1. Velocity profile distortions

With a few notable exceptions, the metrological behaviour of all flow measuring devices is affected to some extent by the velocity distribution at the device inlet section, depending on the construction technology. Table 6.8 summarizes how flow distortions affect each technology.

The best method to reduce the possible errors caused by distortions in the velocity profile is to avoid the causes. In order to achieve an appropriate distribution, a certain length of straight pipe (dependent on the type of distortion and meter) should be installed upstream from the meter. Whenever possible, the cause of the distortion should also be placed downstream from the meter.

However, sometimes it is impossible to eliminate the causes of certain distortions, such as the ones caused by a partial clogging of the filter or the reduction in section

TABLE 6.8. SENSITIVITY TO FLOW PROFILE DISTORTIONS BY METER BY TECHNOLOGY

Instrument	Technology	Sensitivity	Comments
Flow meter	Ultrasonic (transit time)	High	Depends on the number of measuring trajectories and planes
Flow meter	Electromagnetic	Medium	
Flow meter	Insertion probe	Very high	
Water meter	Single jet	Low	Depends on the type of distortion
Water meter	Multiple jet	Very low	
Water meter	Positive displacement	None	
Water meter	Horizontal Woltmann	High	
Water meter	Vertical Woltmann	Low	
Water meter	Paddle wheel	Very high	
Water meter	Proportional	Medium	

FIGURE 6.18. FLOW PROFILE DISTORTION BEFORE A WOLTMANN METER

FIGURE 6.19.   PARTIAL CLOGGING CAUSED BY LIMESCALE AND SUSPENDED PARTICLES

caused by the choking of the joint (Figure 6.19). These situations are quite common in domestic meters and the response of the meter mostly depends on the way the filter is blocked. Error curve tests carried out show the importance of the nature of the blockage, and the meter design.

As a general rule, it can be said that when the blockage concentrates the flow in a single space (Figure 6.19, top right), velocity meters (single jet) are the most affected.

6.6.2.   Installation position of the meter

Quite often, the space available for the installation of a meter is minimal, leading to an incorrect position of the meter, making it almost impossible to read. This problem is not just limited to old households. Many new buildings lack the appropriate facilities to allow a proper installation of the meters. Figure 6.20 shows several situations with inappropriate meter installations.

Meter manufacturers are not helpful in solving the problem. Many of the designed totalizers are quite large and difficult to read. Very few manufacturers have opted for models that would solve the reading problems in small cabinets. Figure 6.21 shows

FIGURE 6.20. INCORRECT METER INSTALLATION

two meter models. The one on the left has the rollers, which display the total volume, on the top part of the meter (as is usually the case) while the one on the right has the cubic meters display on the lower part of the totalizer. The result is that the meter on the right is easier to read from a greater number of angles.

The problem can become worse when the meter is placed inside a reduced space (like a cabinet). Some manufacturers try to solve the reading angle problem with alternative solutions. Figure 6.22 shows the display of a meter at 45° angle (some other models locate the window for the rollers on the side of the totalizer).

However, the problems from an inappropriate installation of a meter do not end with the difficulties to read. If the meter is installed on a certain inclination, the impeller will not rest appropriately and will increase the friction on the medium term. At low flowrates this greater friction will influence the measure.

Generally speaking, meters may lose a metrological class when installed inclined. Class B meters become Class A, and Class C become Class B. However, and depending on the design, the consequences are different. In average, the increase in the initial error due to inappropriate installations can be estimated in 3%. This error will increase on the medium and the long term.

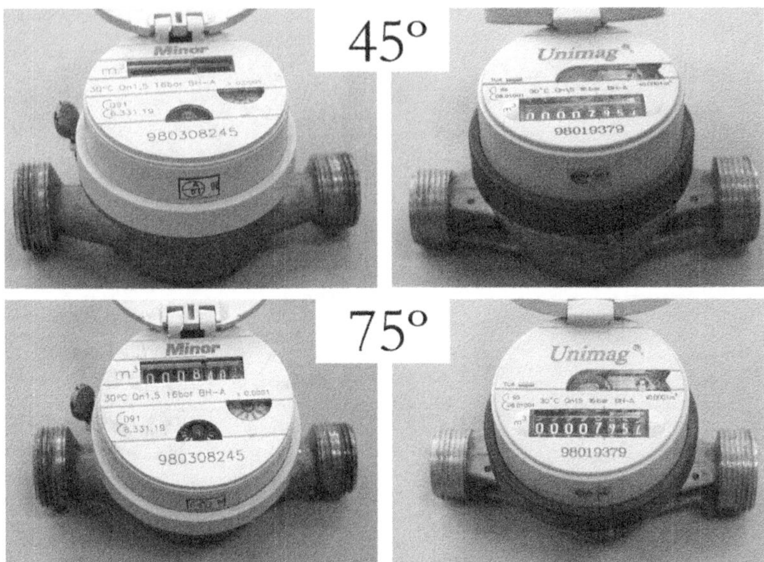

FIGURE 6.21. INFLUENCE OF THE METER'S DESIGN ON THE READING ANGLE

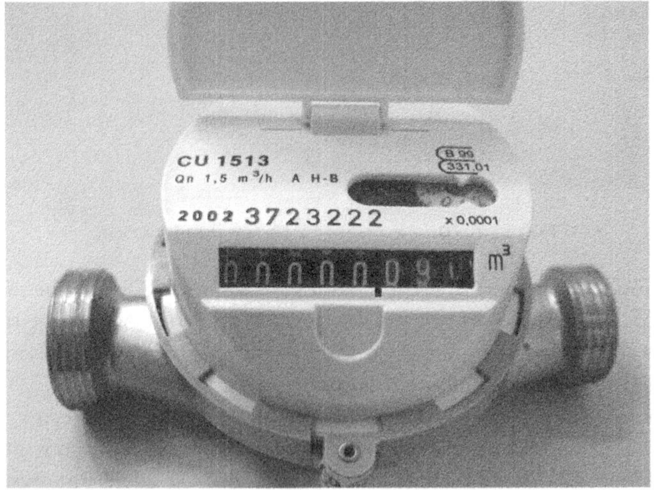

FIGURE 6.22. 45° DISPLAY TO INCREASE READING ANGLE

In order to determine precisely the influence of the installation on the magnitude of the error, several meters were tested against a volumetric Class C meter which served as reference (and was installed properly). A toilet was flushed every 15 minutes a total of 14 times. Table 6.9 shows the error figures (compared with the reference value) that were obtained with the meters installed horizontally and with a 90° rotation. Positive values indicate that the registered volume was higher than the reference value.

TABLE 6.9. INFLUENCE OF INCLINATION IN THE ERROR OF SINGLE JET METERS

	Horizontal position error (%)	90° position error (%)	Difference (%)
Class C	1.3	−6.0	−7.3
Class B (Model 3)	−6.0	−8.0	−2.0
Class B (Model 2)	−7.2	−10.5	−3.3
Class B (Model 1)	−1.2	−7.9	−6.7
Volumetric	0.0	0.0	0.0

Table 6.9 also shows how the Class C model is severely affected by the installation position. The meter in a horizontal position registered 1.3% more than the reference volumetric meter. When it was rotated 90°, the error increased in 7.3% all the way up to −6%. In these tests, the meter that was less affected by the inclination was the Class B Model 3, which only lost a 2% error once it was rotated.

In view of the differences found, it can be concluded that in those utilities where a significant number of meters is installed with a certain inclination, it is recommendable to select models where the inclination does not excessively affect the measure. It is important to point out that there are models available in the market that will retain their metrological class regardless of the position in which they are installed.

However, the concerns related to inappropriate installation positions should not be limited to this initial error. The wear of the moving parts of the meter may be increased in inclined positions, and consequently the rate of decay may also be higher. It is advisable to perform wear tests to try to predict the future behaviour of the meter and the evolution of the error curve in inclined positions.

6.6.3.   Suspended solids and limescale build-up

The error curve of most meters, regardless of their technology, may be seriously affected by the quality of the water. For instance, single jet meters' metrology is closely linked to the internal dimensional tolerances. Additionally, several single and multiple jet meters present a by-pass to adjust the error curve. In these cases, the error is directly related to the amount of water circulating through the by-pass. A reduction in the cross section of the metering chamber may create over-metering. In some models these errors can be quite severe, with the adverse consequences for users. If the by-pass section is reduced, the error curve will be displaced towards positive errors. This is why meters with error curve adjusting devices should be periodically inspected, for an important number of users could be seriously affected.

In some other cases, an excessive build-up of limescale will influence the rotation of the impeller, even stopping it, leading to a significant under-registration (Figure 6.23). Fortunately, some alarm signals appear in these cases, and the billing figures usually begin a gradual reduction before the impeller is completely blocked.

6.6.4.   Fogging

Fogging of the totalizers display is a problem that usually occurs in humid areas when meters are installed outdoors at ground level. The accumulation of condensation in

FIGURE 6.23. LIMESCALE BUILD-UP IN METERS

the display introduces additional difficulties in meter reading, and sometimes it even impedes it. Occasionally fogging will produce wrong readings and originate billing complaints from users.

This problem can be solved, at least in theory, by using hermetically sealed extra-dry totalizer or wet registers. In practice, even the latest models may present fogging after some time. The most effective solution is once again prevention, and meters should be field tested in humid conditions to see which models respond better to condensation problems in the totalizer.

6.6.5. Meter tampering

The manipulation of meters by third parties in order to alter their accuracy will obviously create problems in the measurement of consumption. There are many techniques used to tamper the meters. Quite obviously, as new fraud preventing systems are developed, new methods are created to neutralize them. Some of the most common manipulation attempts are:

- Introduction of a needle or similar object in the totalizer to block the motion of the gears. This element is periodically removed as the reading date approaches. Sometimes the foreign object is simply introduced before an important consumption is made (filling a pool, watering the garden).

FIGURE 6.24. TAMPERED METER

FIGURE 6.25. WARNING MECHANISM TO DETECT IMPACTS ON THE METERS TOTALIZER

- Introduction of foreign objects in the metering chamber to block the impeller. In order to perform this operation the meter needs to be removed from the installation. If precincts are used at the time of installing the meter, the manipulation will be easily detected (Figure 6.24).
- Placement of powerful magnets to break the magnetic coupling of impeller and gears. New models incorporate protection rings which make this operation more difficult.
- Impacting strongly the meter at a certain point. Quite often, some models will significantly increase their error after such impact. Some models incorporate mechanisms to warn about this type of fraud (Figure 6.25).

FIGURE 6.26.   TAMPERING OF A METER BY OBSTRUCTION OF THE OUTLET OF THE
METERING CHAMBER

- Temporary removal of the meter. Although it is quite easy to detect, often utilities will have problems proving that it was the user who altered the integrity of the meter.

Figure 6.26 shows a failed attempt of tampering a meter. Once the meter was tested, it was shown that the error curve had not been significantly altered. The manipulation tried to partially reduce the outlet of the meter in an attempt to reduce the rotation speed of the turbine. This manipulation was detected in a utility in a high number of meters (4 out of a sample of 200 meters). It was concluded that it probably originated from personnel specialized in installations or plumbing.

The extended use of electronic meters in the future will change completely the ways in which meters are currently tampered.

6.6.6.   Meter size

The selection of the right size of meter is not especially important in small calibre meters (with the possible exception of households with pool and garden). However, in medium and large calibre meters, sizing is a key factor for an adequate quality of metering. A significant proportion of the metering errors in large users is due to a wrong selection in the size of the meter. On one hand, if the meter is working at flowrates which are too low, the under-metering will be significant. On the other hand, if the meter is subject to flowrates which are too high, the decay of the error curve may be accelerated, leading to important errors in a relatively short period of time after the installation.

In order to determine the right size of a meter, and guarantee that the working range of flowrates is adequate, three main methods can be used:

1. When the study refers mainly to domestic users, the peak flowrate can be estimated from the installed flowrate. This method is often used to determine peak flowrates in installations inside buildings (for instance taking into account the number of appliances and a possible simultaneity in the uses).

TABLE 6.10. PEAK FLOWRATES ESTIMATION BY TYPE OF HOUSEHOLD

Typically installed flowrate (l/s)	Number of appliances	$K = 1/(j - 1)^{1/2}$	Expected peak flowrate for the household (l/s)
0.40	3	0.71	0.28
0.80	5	0.50	0.40
1.40	9	0.35	0.49
1.95	12	0.30	0.59
2.30	15	0.27	0.62

Being more specific, the method will add for each household the nominal flowrates for each consumption points, and then divide them by the square root of the number of devices minus one. This operation provides an estimation of the peak flowrate for each type of household (Table 6.10).

The peak flowrate that would originate in N households of the same type would be calculated with the following expression:

$$Q_{N\,household} = \frac{19 + N}{10(N + 1)} \cdot N \cdot Q_{peak\,household} \tag{6.7}$$

$$Q_{peak\,household} = k \times Q_{installed} \tag{6.8}$$

2. Another commonly used method takes into account the expected monthly consumed volume to select the nominal flowrate of the meter.

   Even though this approximation is not as direct as the previous one, it is often used due to the availability of consumption data vs. the possibility of obtaining reliable data on the number of appliances and internal facilities of the users. This method is recommended by some meter manufacturers, since it allows to evaluate by means of the accumulated registered volume and the age of the meter whether it is subject to higher or lower flowrates than the recommended ones.

   In order to estimate the consumption of a certain sector, and assuming that no real data on the use of water are available, it is possible to use tables like Table 6.11, taken from the Brazilian Standard NBR 5626/82. From the consumption data and with the information provided in Table 6.12, which presents the recommendations of the manufacturer, it is possible to select the nominal flowrate of the meter to be installed.

3. The last method is the most precise one and is based on determining the consumption profile in order to accurately establish the flowrates circulating through the pipe. This profile can be either obtained with the original meter or by replacing it with a more accurate one. If the original meter is kept, it is crucial to correct the obtained data with the possible bias that the meter may be introducing in the obtained profile. For instance, if a meter has deteriorated and is not replaced when calculating the profile, the obtained data will not be reliable. In this case, the absence of consumption at low flowrates will not necessarily mean it does not exist, but rather than the meter is not able to register due to lack in sensitivity at the lower range of flowrates.

TABLE 6.11. ASSIGNATION OF FLOWRATES TO DIFFERENT USES

Estimated consumption of the installations (l/day)	
Residential	150 per capita
Popular or rural houses	120 per capita
Apartments	200 per capita
Provisional lodgings	80 per capita
Restaurants and similar	25 per meal
Hotels (no kitchen or laundry service)	120 per guest
Hotels (with kitchen or laundry service)	250 per guest
Schools – external	50 per capita
Cinemas and theatres	2 per seat
Laundry	50 per kg of dry clothes
Temples	2 per seat
Orphanages and asylums	150 per capita
Markets	5 per m^2
Commercial buildings	50 per capita
Car washes	150 per vehicle
Hospitals	150 per bed
Factories (only personnel consumption)	70 per employee

TABLE 6.12. SELECTION OF THE METER'S NOMINAL FLOWRATE AS A FUNCTION OF CONSUMPTION

Estimated consumption		Nominal flowrate (m^3/h)	Example of an adequate meter
(m^3/month)	(m^3/day)		
0–180	0–6	1.5	Single jet 15 mm
120–250	4–8	2.5	Single jet 20 mm
210–350	6–12	3.5	Single jet 25 mm
300–540	8–18	5	Single jet 32 mm
430–900	14–30	10	Single jet 40 mm
750–1500	25–50	15	Single jet 50 mm
1200–4500	40–120	25	Woltmann 50 mm
1800–7500	90–250	40	Woltmann 65 mm
4500–13000	180–500	60	Woltmann 80 mm

Example: Selection of a hotel meter

The water distribution facilities of hotels may be very different depending on each case. For this reason, it is almost impossible to develop standard rules for meter size selection. In this example, the hotel had a storage tank from where most appliances were supplied. The hotel also had a direct connection for the swimming pool. Generally speaking, the flowrates dedicated to fill the tank were very small. However the flowrate used to fill the pool, with negligible friction losses in the installation, was quite high.

The four star hotel had 196 rooms and 3 suites. Additionally the hotel had a restaurant, a bar, a gym and three meeting rooms. The water consumption of the hotel was

TABLE **6.13.** SELECTION OF THE NOMINAL FLOWRATE OF A METER AS A FUNCTION OF
CONSUMPTION

Range of flowrates (m^3/h)	Used volume (l)	Percentage of consumption
0.100	0	0
0.1–0.3	92,280	7
0.3–1.2	473,570	36
1.2–2.5	310,350	24
2.5–5.0	235,520	18
5.0–10.0	162,370	12
10.0–20.0	27,920	2
20.0–40.0	0	0
>40.0	0	0

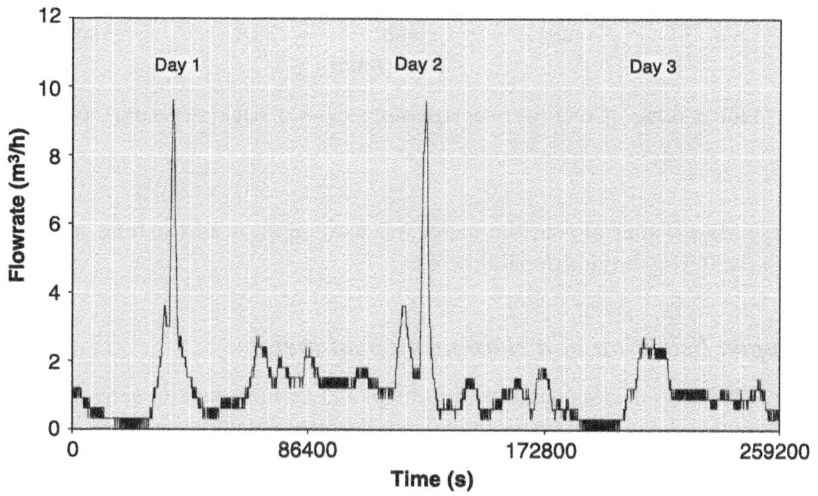

FIGURE **6.27.** TYPICAL CONSUMPTION FLOWRATES OF THE HOTEL FOR 3 DAYS

measured for 51 days and presented an average water consumption of 25 m^3/day. The distribution of the consumption flowrates is shown in Table 6.13, while Figure 6.27 shows the consumption in 3 typical days. It can be seen that flowrates passing through the meter are usually below 5 m^3/h, corresponding to the maximum flowrate of a 20 mm meter.

From the information collected to elaborate the water consumption pattern, the utility should install a single jet Class C meter, of 32 mm (Q_n 6 m^3/h) or 40 mm (Q_n 10 m^3/h). The first one would occasionally work (2 h in 51 days, 0.16% of the time) close to its maximum flowrate. The second option would provide a wider margin.

If the same selection had been done by taking into account the average consumption, 25 m^3, the most adequate meter according to Table 6.12 would have been a single jet 40 mm meter.

The actual meter selected by the water utility in this case had been a single jet meter of 65 mm, with a nominal flowrate of 20 m^3/h, which in practice was clearly

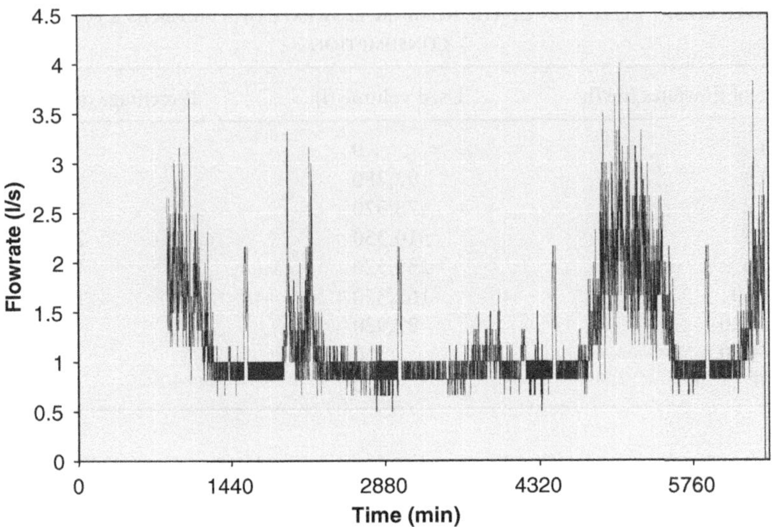

FIGURE 6.28. CONSUMPTION FLOWRATES IN AN OLYMPIC SWIMMING POOL

too large. As a matter of fact, the meter was working most of the time at flowrates close to the 0.3 m^3/h transitional flowrate.

Example: Selection of a swimming pool meter

Figure 6.28 shows the average consumption of an Olympic pool for several days. The meter found at the installation was an 80 mm horizontal Woltmann meter. This meter has a nominal flowrate of 60 m^3/h (16.67 l/s) and a peak flowrate of 120 m^3/h (33.3 l/s).

Quite obviously the existent meter was too large for the job, and a 50 mm Woltmann meter would have been more than enough. However, in this case the best option would be the installation of a 50 mm single jet Class C meter, with a nominal flowrate of 15 m^3/h (4.17 l/s) or even a 40 mm one with a nominal flowrate of 10 m^3/h (2.78 l/s).

Another conclusion that can be obtained from the figure is that the second method presented to determine the size of the meter would have provided in this case almost an identical meter. The daily consumption of the pool was approximately 100 m^3/day, maybe a little oversized. According to Table 6.12 the most adequate meter would be a 50 mm Woltmann with a nominal flowrate of 25 m^3/h.

In summary, the determination of the flowrate profile is a very useful tool for determining the appropriate size of a meter. The largest the pipe diameter, the more important it is to actually know the real consumption flowrates in order to adequately select the meter. In these cases, the configuration of the internal installation plays a critical role in the real flowrates. The alternative techniques available to determine the meter size usually lead to similar results. However, the uncertainty of the adequateness of the choice is always higher than when using the consumption pattern.

FIGURE 6.29. STORAGE TANKS ON HOUSEHOLD ROOFS

6.6.7. Storage tanks in households

In several countries, the use of storage tanks in households is quite extensive due to several reasons (ranging from tradition, to frequent service interruptions and including network insufficiency to satisfy demand) (Figure 6.29).

Regardless of many other considerations which may deem inconvenient the use of such tanks, these storage tanks have a considerable economic impact on the utility's revenue. In our experience, the flowrates circulating through the pipes of such buildings are much lower than average. This causes metering errors close to -20%. The fact that average uses will only provoke slight changes in the tank level causes that the entry flap valve will only admit very small flowrates. The solution to this problem lies in changing the valve type to an instantaneous closure admission valve.

6.6.8. Installation of meters inside the households

Block tariffs have become more and more frequent in utilities worldwide. However, in order to effectively apply tariffs that penalize higher consumption, the service provider needs to be able to perform regular and periodic readings of all meters in the utility. In practice, many companies estimate the consumption of many users due to the impossibility to perform actual readings of the meters during the billing period of time. Even though the company may have the human resources to perform this task, sometimes the meters are located inside the households and are not accessible by the utility's staff.

The location of meters inside the households creates additional costs, not only related to the higher difficulty in reading them, but also related to their replacement and the non-revenue water flowing through them. Replacing a meter inside a household requires an appointment with the user and in many occasions the replacement of meters has to be done individually (whereas other meters can be replaced in bulk). For this reason, and although a meter inside a household may be deteriorated or stopped, the utility

FIGURE 6.30.   METER INSIDE A HOUSEHOLD IN FRANCE

may choose to maintain the estimation of consumptions rather than facing the difficulties of replacing it. As a matter of fact, quite often the average error of meters inside households is much higher than the one corresponding to those meters easily accessible by utility's staff.

In those cases where the meter is installed inside the households, it is recommended to install higher-quality models which are able to maintain their metrological characteristics for longer.

7

Maintenance and renewal of meters in a utility

7.1. INTRODUCTION

The installation and periodic collection of data from water meters allows utilities to know the volume of water consumed by every user and bill accordingly with that information. Water meters are consequently a cornerstone in the management of a water utility. Choosing the right meter model and maintaining and renewing the meters properly will allow to maximize revenue and reduce costs.

It should be pointed out that *properly* maintaining the meters does not necessarily mean that all meters should be maintained and renewed to minimize their error. On the contrary, for every utility there will be an optimum replacement frequency and an adequate maintenance policy that will fulfil the economic objectives described above. This replacement frequency will obviously depend on the water tariff, the cost of the meters and the type of users.

As a result, there are two main choices to be made. The first choice will be to select the right type of meter. In this case, it will be a matter of considering its metrological quality and metering technology, its measuring range, durability, acquisition cost, etc. The second choice will be to determine how long those meters should remain installed to optimize their use. This last choice may also be conditioned by additional variables, such as local regulations that may require a certain replacement frequency. In any case, both questions cannot be answered separately.

The solution of the problem is mainly of an economic nature. With independence of the additional restrictions that may need to be applied, the right solution is the one that fits better the utility's policy in terms of investment and revenue, usually maximizing benefit (revenue minus investment). This implies a cost-benefit analysis for every meter model which will be developed later in the chapter.

7.2. RESTRICTIONS

The key decisions regarding meters in a utility are often conditioned by restrictions and local regulations that must be taken into account before making a final choice. These are the most common ones.

7.2.1. Regulations on the metrological quality of the meters

Most meter manufacturers only provide Classes B and C (ISO 4064:1993) meters for urban use nowadays. Class A meters are almost never used in developed countries. Class D meters are extremely expensive and sensitive to foreign particles in the water. They are rarely used, and although they can sometimes be found in countries like the UK, their use can only be justified by a high price of water. The choice, regarding the metrological class, is consequently limited mostly to Class B or C. Obviously, new standards, already published, will change this scenery since the metrological categories that they define are more numerous.

Whenever there is a legal restriction on the minimum metrological class eligible, the economic analysis will only consider those models complying with the regulation. Although most manufacturers will only offer a model or two per metrological class, the differences between manufacturers can be quite significant, and so will be the consequences of the choice made.

It is important to point out that the new standards, which have been recently published, consider a larger number of categories, and take into account the value of the flowrate at which the meters begin to perform accurately according to the maximum permissible error. This reflects the claim of many manufacturers which have often alleged that not all meters of the same metrological class have a comparable error curve (Figure 7.1).

In any case, and as important as the initial error curve is, the final decision should also take into account how this curve evolves with time (as described in Chapter 6).

One final note must address the situation of those countries following the AWWA standards. These standards set error limits and measuring ranges for every diameter, and do not use the concept of metrological class. In this case, there will be no restrictions on the metrological class to be used.

7.2.2. Regulations on the renovation frequency

Quite often, due to national or local regulations, meters must be replaced after a certain period of time. Some other regulations may limit the total registered volume as a function of the nominal flowrate of the meter. This later form of replacement policy is better in many cases where the installation date of the meter is not known with certainty.

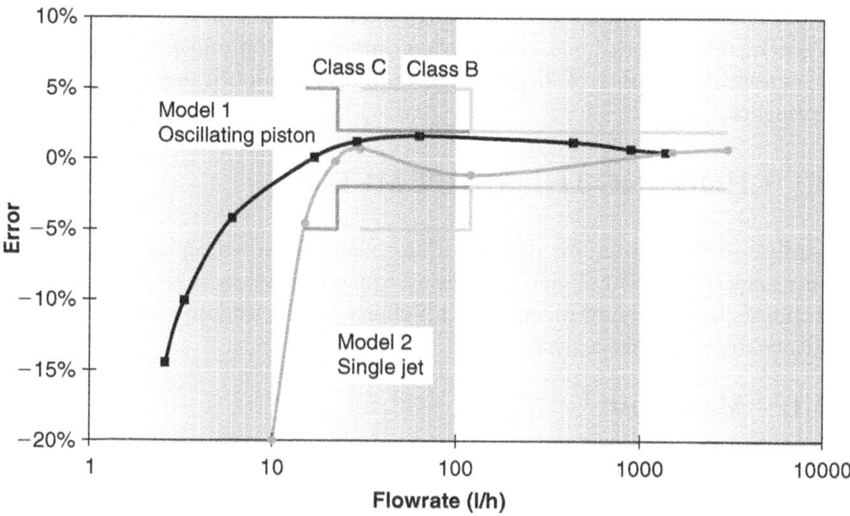

FIGURE 7.1. ERROR CURVES OF TWO CLASS C METERS OF 1.5 M^3/H NOMINAL FLOWRATE

These type of regulations usually offer an alternative for the utility. The instrument can be replaced or it may be maintained in operation after being calibrated either in situ or at a laboratory. The low price of mechanical meters nowadays often deem the first option the best (although the calibration costs may be considered part of the costs associated with the decision making model).

Applying these regulations properly requires the use of an updated database with data on the users and their meters. Currently, and depending on the part of the world, the number of utilities with such a database is not too high or the data are not too reliable. Setting up and maintaining these databases up to date is expensive and many times the price of water does not justify it.

7.2.3. Renovation based on the maximum global error

In other occasions, the utility wishes to limit the weighted error of the meters. This guarantees a certain equity in the water bill among users. It must be taken into account that as meters age, the dispersion of the errors increases, and as a consequence some users may be paying other users' water (or at least their share of the costs).

7.2.4. Renovation based on basic maintenance criteria

The utility may also choose to carry out basic maintenance on the meters, replacing only those that are suspicious of being clogged or are considerably old or used (high volume). This policy is only admissible in those utilities with a low price of water and metering costs must be kept very low, or exceptionally in those households where accessing the meter becomes a problem (i.e. when located inside a household). In these cases, the number of visits necessary to replace the meter becomes a key component of the replacement costs.

The results of this maintenance policy will lead to significant metering errors and, as a general rule, to a mediocre control of consumption. Additionally, the low control will increase the number of illegal connections and manipulations of the metering instruments.

7.3. CHOOSING THE RIGHT METER

Regardless of the restrictions that the utility may find when replacing the meters, from a managerial point of view it is always convenient to assess all possible options and choose the most convenient one. This choice will depend on several factors that influence the economic analysis.

7.3.1. Meter cost

The cost of a meter includes both its price and the cost of installation (including administrative costs). The quality of a meter is usually related to its cost, and it is reasonable to expect that a cheaper meter will deteriorate faster than a more expensive, better built meter. Depending on other variables, like the price of water, this cost will be recoverable sooner or later (or it may not be recoverable during the lifespan of the meter).

A cheaper price of water will usually lead to the installation of cheaper class B meters. On the other hand, a higher water tariff will advise to install higher-quality meters.

7.3.2. Error curve

Besides the costs associated to the meter itself, some other costs should also be considered: those related to the economic losses of non-metered water. These costs depend on the capacity of the meter to register all water consumption (i.e. its error curve).

It has already been shown that the error curve of a meter is related to the metering technology and the design of the instrument. In any case, the quality of the meter is clearly shown in its sensitivity at low flowrates, and the capacity to maintain it with time. In this range of flowrates, the only means that the manufacturer has in order to improve the meter's performance are a better design and a higher manufacturing quality.

When new meters are tested and their weighted errors are calculated, the differences within the same metrological class are not significant. The shape of the error curve usually accounts for differences in the weighted error and they usually never differ in more than 1% or 2%.

Only with the lapse of time the higher-quality models will stand out noticeably. Meters made with lower-quality materials and rougher tolerances will lose accuracy faster and differences between models will become significant. In those situations where the price of water is high, it might pay off to install the meter that preserves better the error curve.

However, additional factors may need consideration when choosing a higher-class meter. The tighter tolerances and the great reduction of friction between the moving parts mean that these meters may be subject to greater wear in the presence of a water of poor quality. This explains the many cases where some of the most sensitive meters have been found blocked or stopped, or have lost their sensitivity at low flows, after a short period of time.

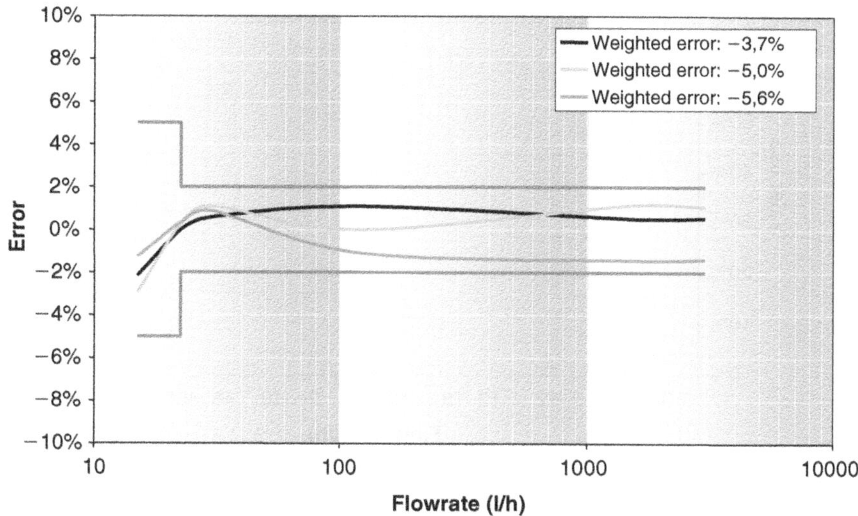

Figure 7.2. Error curves of three different Class C single jet meters

7.3.3. Consumption pattern

The total amount of water not registered by a meter depends entirely on the consumption pattern of the user. The same meter installed for two different users may present completely different errors.

For instance, Figure 7.3 presents the consumption flowrates of two very different users. The first one presents *normal* consumption flowrates, while the biggest consumption of the second one corresponds to a leak with a flowrate below 6 l/h. If Model 2 meter of Figure 7.1 was to be used for both users, metering errors would be completely different: for the first user, the weighted error would be of −6%, while it would be reach values above −80% for the second user.

It may be stated that the biggest influence in the weighted error comes from consumptions at the lowest flowrates, where meters have a worse performance and appear the greater changes in the error curve. Although many experts do not completely agree with that statement (arguing that the consumption at these flowrates is not very large in volume) they may not be taking into account three important facts:

1. Although the percentage in the total consumption is low, the magnitude of the error (100% below the starting flowrate) compensates this.
2. The error curve at medium and high flowrates does not deteriorate almost at all with time. This will in most cases maintain the average error for nominal flowrates within the ±2% range.
3. The error curve deteriorates very quickly at low flowrates. As a consequence the weighted error increases very quickly with time when consumptions are concentrated in this range of flowrates.

A direct consequence is that a consumption pattern with a high percentage of low flowrates will influence on the initial value of accuracy, but it will also reduce the lifespan

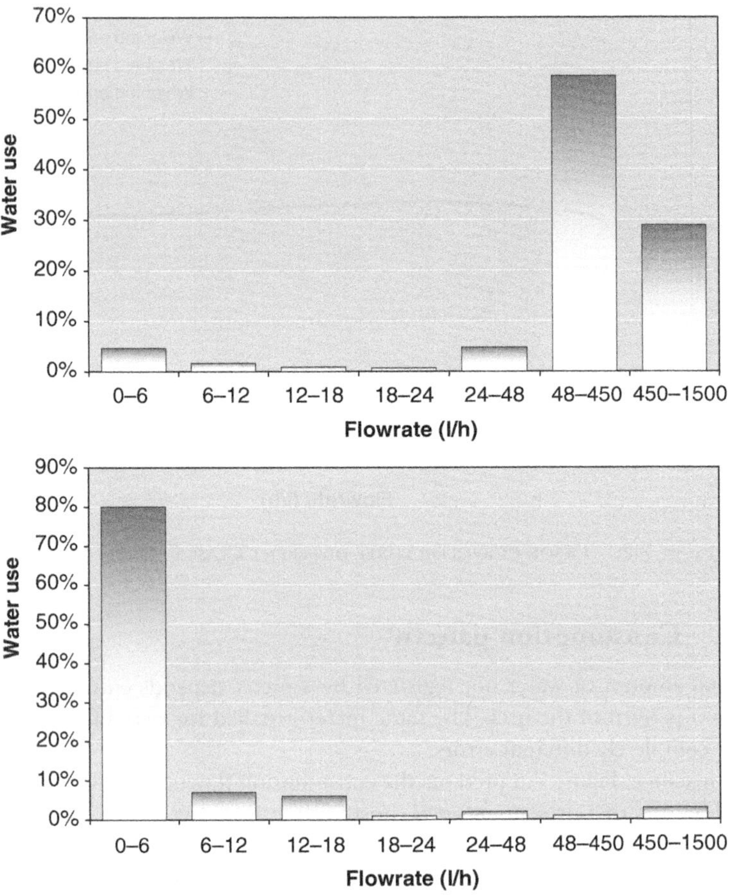

FIGURE 7.3. DIFFERENT CONSUMPTION PATTERNS IN HOUSEHOLDS

of a meter (with a certain level of operation). For instance, a user with demands over 100 l/h, where the error curve does not change with time, will allow the meter to remain in operation for many years without having the non-metered water change significantly. On the other hand, a user with leaks will have its registered consumption changed the minute there is a shift in the error curve.

7.3.4. Water tariff

The economic losses derived from non-metered water are proportional to its selling price. From another point of view, the revenue generated by a specific meter depends on the registered volume and the price of water. It is obvious that the water tariff plays a key role in the economic analysis of meter management.

It is important to properly assess the impact of water price in this economic analysis. For example, in those utilities where the water tariff is structured in blocks, the loss in revenue from under-metering will have to be assessed at the price of the last cubic meter registered by the meter. Should the meter have registered the additional consumption, the water bill would have been increased using the price of this last block in the tariff structure.

With increasing water prices, the need for efficient management of the installed water meters becomes more important. In situations with high water prices, the renovation policy must be effective and on time, for a 2–3 years miscalculation in the optimum renovation period may lead to a significant loss of revenue. When the price of water is too low, an inaccuracy in the renovation policy will have a much lesser impact on the water company incomes.

Consequently, with expensive water prices, the renewal frequency will need to be higher. The chances to use higher-quality meters will increase and the estimation of the renovation period will become a key decision that will influence future revenues.

7.3.5. Discount rate

The nominal discount rate, r, is the parameter used to obtain the actual value at present time of future economic figures. In the case of water meters, since the investment is reasonably risk free, the discount rate used will be quite close to the ones set by each country as risk free rates (i.e. state bonds).

It is necessary afterwards to deduct from this nominal rate the effects of inflation s. The resulting figure will be the real discount rate r', which can be calculated by:

$$r' = \frac{(1 + r)}{(1 + s)} - 1 \tag{7.1}$$

The greater r' is, the lesser will be the value of future economic amounts translated into today's money (present value). For instance, if a certain economic amount P_i is to be accounted in the year i, on the future life of the meter, its present value, to be comparable with today's amounts, would be calculated as shown by:

$$P_{\text{actual}} = \frac{P_i}{(1 + r')^i} \tag{7.2}$$

Just to further illustrate this concept, Table 7.1 shows the value of one monetary unit after a number of years (rows) and as a function of the discount rate (columns). If the real discount rate is 0%, the monetary unit retains its value through time. For a discount rate of 8%, that same unit in 20 years time after the beginning of the project will only be worth in today's currency approximately one fifth of its numerical value (0.21).

The discount rate value depends on the economic objectives of the utility. Selecting a high-discount rate means that the utility expects high revenue from the investment, and it restricts the possibilities of using higher-quality meters, which are more expensive. In this case, it becomes more difficult to recover the initial investment since the losses avoided in the future will not be worth much due to the high-discount rate chosen, and consequently the increased future accuracy is valued less than the initial investment.

TABLE 7.1. PRESENT VALUE OF ONE CURRENCY UNIT FOR DIFFERENT REAL DISCOUNT RATES

Years	Real discount rate (%)				
	0	2	4	6	8
0	1.00	1.00	1.00	1.00	1.00
2	1.00	0.96	0.92	0.89	0.86
4	1.00	0.92	0.85	0.79	0.74
6	1.00	0.89	0.79	0.70	0.63
8	1.00	0.85	0.73	0.63	0.54
10	1.00	0.82	0.68	0.56	0.46
12	1.00	0.79	0.62	0.50	0.40
14	1.00	0.76	0.58	0.44	0.34
16	1.00	0.73	0.53	0.39	0.29
18	1.00	0.70	0.49	0.35	0.25
20	1.00	0.67	0.46	0.31	0.21

7.4. ECONOMIC STUDY

The main characteristic of water meters is that they represent renewal or replacement investments. The utility needs to replace them once they have covered their lifespan so the business activity can continue. The two most used procedures to select this type of equipment goods are the net present value (NPV) and the equivalent annual annuity (EAA).

NPV consists in estimating the different updated cash flows of the project and from this figure, the initial investment is deducted. A positive quantity would mean that the investment is profitable and feasible. Another way of assessment (when the question is not whether to carry out the project, but rather how) would consist in calculating the updated costs and seek the least cost option.

The more or less simplified methodologies that have been developed to obtain the optimum meter model for a utility, and its renovation frequency are based in this last procedure. In all methodologies (e.g. Allender, 1996; Arregui, 1999; Male et al., 1985; Planells et al., 1987; Taborda and Anunciaçao, 1997; Yee, 1999) the proposed models rely on the evaluation of the total cost of the instrument for the utility during its lifespan.

These costs can be grouped in two categories as shown by:

$$\text{Costs} = C_{\text{purchase}} + C_{\text{inst}} + C_{\text{adm}} + \sum_{i=1}^{n} \forall_i \cdot \varepsilon_i \frac{C_{H_2O}}{(1 + r')^{(i-1)}} \tag{7.3}$$

where:

C_{H_2O} is the water price.

$\forall_i$ is the average volume consumed by a user on the year i.

ε_i is the weighted metering error of the meter on the year i.
r' is the real discount rate.

The fist category would include purchase, installation, administrative and maintenance costs. Usually in small meters maintenance costs do not exist and only the first three must be considered.

On the other hand, the costs originated in metering errors must be considered. These costs usually increase with time, and its rate of increase depends on the characteristics of the utility and the meter (in some cases they may even decrease with time). These costs must be corrected with the real discount rate in order to compare them with current costs.

If the focus is placed on the revenues instead of the costs, the procedure would be similar and only the costs of non-metered water and the signs in Equation (7.3) would change (Equation (7.4)). In any case both methods are equivalent:

$$\text{Revenues} = -C_{\text{purchase}} - C_{\text{inst}} - C_{\text{adm}} + \sum_{i=1}^{n} \forall_i \cdot (1 - \varepsilon_i) \frac{C_{\text{H}_2\text{O}}}{(1 + r')^{(i-1)}} \tag{7.4}$$

The main difficulty in applying Equations (7.3) and (7.4) is to assess the rate of decay of the weighted metering error, allowing to calculate the revenues obtained with each model as a function of time. This topic was already covered in Chapter 6. The problem is that this information will usually not be available a priori when first selecting a meter. However, once installed, it will be possible to analyze the error curve of a selected sample of meters and calculate the optimum renovation frequency with the actual behaviour of the instruments.

7.4.1. Selection of the most economic meter for a specific renovation frequency

If the renovation frequency is fixed for whatever reason, choosing the most appropriate meter becomes relatively easy. The best model will be the one that maximizes revenue (minimizes costs) according to Equations (7.3) and (7.4). Such meter will maximize the benefits for the utility given the previous constraints.

It is important to point out once more that there are uncertainties related to the estimation of future revenues. Knowing the accuracy and reliability of the data used in the model for these estimations will be a key factor in assessing the possible miscalculations. This is especially important when determining the error rate of decay of the meter, which is usually an unknown factor when purchasing the meter. This problem should be solved by performing a sensitivity analysis that will present different possible scenarios and aid in the decision making process.

Example
A utility needs to buy new meters. The local regulations establish an obligatory renovation frequency and meters must be replaced every 10 years. There are no restrictions on the metrological class. The choice should be made between the options shown by Table 7.2.

TABLE 7.2.   OPTIONS FOR A METER RENOVATION PROGRAMME

	Model A	Model B	Model C
Retail price (€)	15	20	25
Installation costs (€)	12	12	12
Administrative costs (€)	2	2	2
Metrological class	B	B	C
Initial weighted error (%)	−6	−5	−4
Weighted error rate of decay (%/year)*	−0.3	−0.3	−0.3

*Estimated.

TABLE 7.3.   NPV OF THE INVESTMENT FOR DIFFERENT REPLACEMENT FREQUENCIES

Year	Initial costs (€)	Registered volume (m^3)	Present value of annual revenues from registered water (€)	Present value of cumulative revenues from registered water (€)	Total NPV of revenues (€)
1	−29	112.8	33.84	33.84	4.84
2		112.44	32.77	66.61	37.61
3		112.08	31.73	98.34	69.34
4		111.72	30.72	129.06	100.06
5		111.36	29.75	158.81	129.81
6		111	28.81	187.62	158.62
7		110.64	27.89	215.51	186.51
8		110.28	27.01	242.52	213.52
9		109.92	26.15	268.67	239.67
10		109.56	25.32	293.99	**264.99**

The following parameters define the utility:

Average annual consumption per user:	$120\,m^3$
Nominal discount rate:	5%
Inflation rate:	2%
Real discount rate:	2.9%
Water price:	$0.3€/m^3$

From Equation (7.4), the NPV of the revenues series for the particular case of Model A after 10 years is 264.99€ (Table 7.3).

The NPV of the other two models after 10 years are 263.17€ for Model B and 261.33€ for Model C. Consequently, Model A will provide the greatest revenues in the considered period. However, there are evident uncertainties present in some parameters used for the calculations. For instance, the same weighted error rate of decay

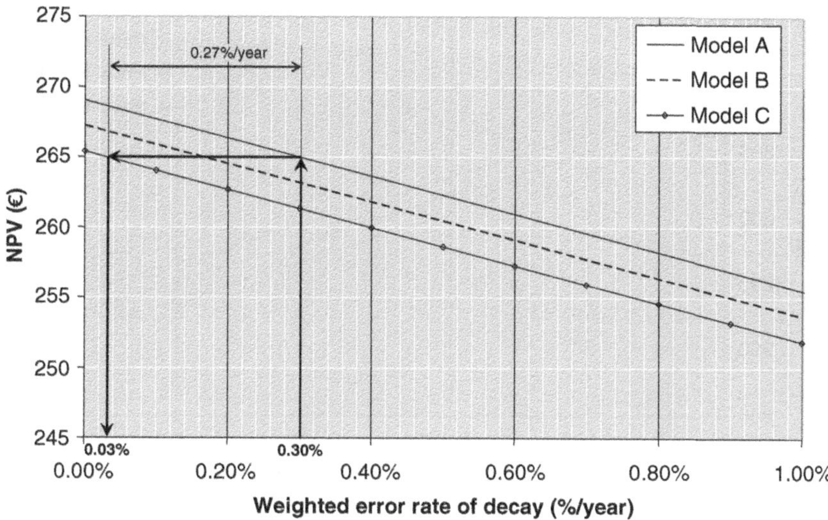

FIGURE 7.4.   NET REVENUE VS. ERROR RATE OF DECAY

has been supposed for all meters. What would happen if this rate of decay was differ-ent for all three models? What should be the rate of decay of Model C to become more profitable than Model A?

In order to give answers to these questions, the following graphs can be made (showing the NPV of each meter for different rates of decay) to assess all possible scen-arios. More specifically and for the data shown in the example, it can be stated that the accuracy of Model C would have to decay at a rate of 0.03%/year (Figure 7.4) to deem this model more profitable than Model A. Regardless of the rate of decay of Model C, Model A would degrade a 0.27%/year faster to compensate the initial price difference between both meters.

In order to assess the influence of several parameters on the outcome of the selec-tion process, the previous figure could be reconsidered with a discount rate equal to the inflation (2%). In this case, the expected return from the initial investment would be zero, and the non-metered water would have a much greater influence in the choice. This would level all choices and instead of 0.27%/year, meter A would only have a 0.17%/year advantage with Model C (Figure 7.5). This could, as a matter of fact, influence the decision, since the difference is not too large and without real data, Model C could be a better choice given its expected better quality.

The price of water has a similar influence in the choice. If the water tariff is increased from 0.3€ to 0.5€/m^3, the value of the non-metered water increases. In such cases, the initial investment becomes almost irrelevant and the required rate of decay for all three models would be almost the same (Figure 7.6).

In summary, the choice of the most adequate meter will depend on several factors, although the price of water is the most decisive one. In any case, the procedure to be used will be as simple as obtaining the NPV of every meter for the desired period and selecting the one that produces larger benefits.

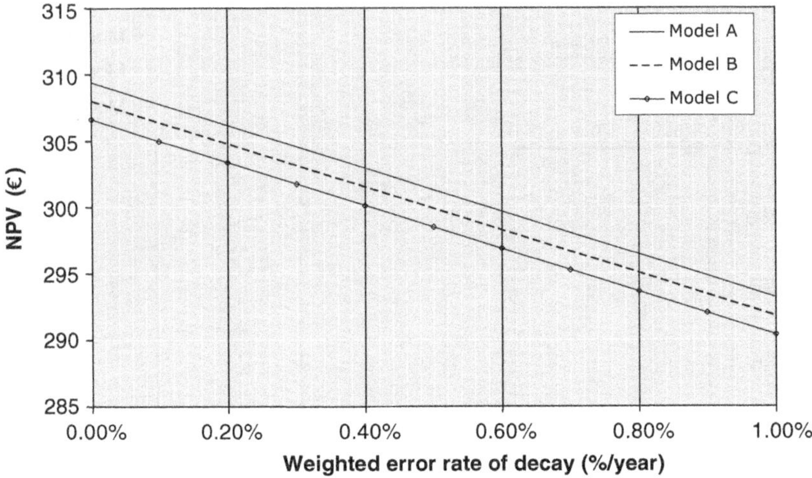

FIGURE 7.5.   NET REVENUE VS. ERROR RATE OF DECAY WITH A 2% DISCOUNT RATE

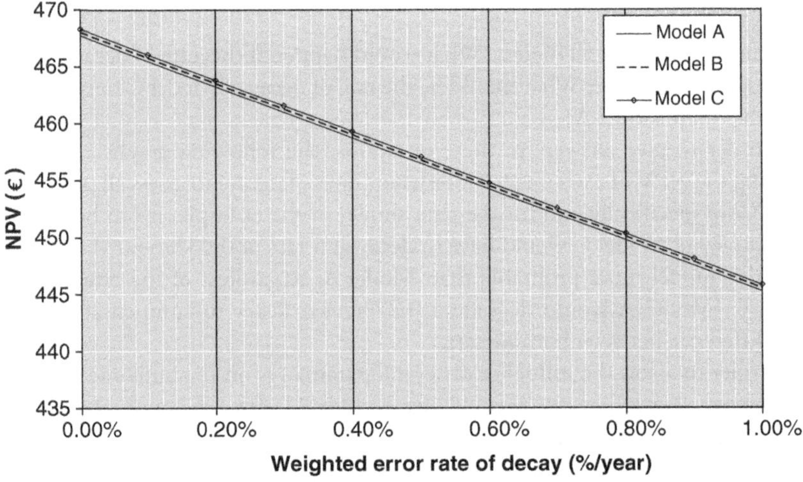

FIGURE 7.6.   NET REVENUE VS. ERROR RATE OF DECAY FOR A PRICE OF WATER OF $0.5€/\text{M}^3$

7.4.2. Selection of the most economic meter without an imposed renovation frequency

A fixed renovation frequency limits the lifespan of all possible meters to a certain value. If this renovation frequency is not imposed, the optimum life of a meter will be the one that maximizes the benefits for the utility. This implies that different models must be compared with different renovation frequencies in order to determine the best investment choice. However, to do this, the simple NPV procedure is no longer an appropriate method and the NPV of the replacement chain (NPVC) must be used.

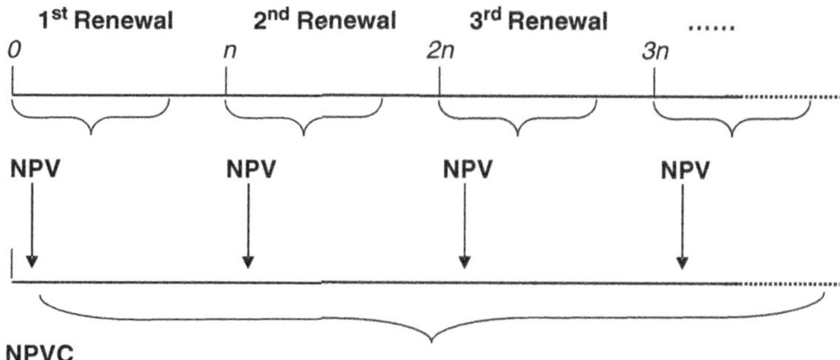

FIGURE 7.7.   NPVC

The NPVC consists in obtaining the NPV of an infinite succession of replacements of the meters (Figure 7.7), and it can be mathematically calculated as shown by Equation (7.6), where r' is the real rate of return:

$$\text{VANC} = \text{VAN} = \frac{\text{VAN}}{(1+r')^n} + \frac{\text{VAN}}{(1+r')^{2n}} + \frac{\text{VAN}}{(1+r')^{3n}} + K$$

$$= \text{VAN} \cdot \left[1 + \frac{1}{(1+r')^n} + \frac{1}{(1+r')^{2n}} + \frac{1}{(1+r')^{3n}} + K \right] \qquad (7.5)$$

$$\text{VANC} = \frac{(1+r')^n}{(1+r')^n - 1} \cdot \text{VAN} = m \cdot \text{VAN} \qquad (7.6)$$

In order to determine the lifespan of a given meter, the NPVC for every alternative is determined. The optimum renovation frequency will be the one to generate larger revenues (or minimum costs). Consequently, the optimum lifespan of the meter will be determined by comparison of the NPVCs.

Under no circumstances should the simpler NPV calculation be used to search for a minimum in the function and choose the meter. This procedure, although is proposed often in the literature (i.e. Allender, 1996; Yee, 1999) does not provide the least cost option and lead to miscalculations.

Example
The utility of the previous example no longer has a restriction on the renovation frequency. All other variables remain the same.

In such a case, two main problems need to be solved. First the most adequate model should be selected. Afterwards, the optimum renovation frequency should be calculated. And the answer to both questions must be found simultaneously. In order to calculate the NPVC (last column of Table 7.4) the NPV of each renovation frequency must be multiplied by the coefficient m (Equation (7.5)).

Table 7.4 shows that for Model A, the optimum renovation frequency is 26 years and the associated NPVC would be 1088.5€. Final results for all three models are shown in Table 7.5.

TABLE 7.4. NPVC OF THE INVESTMENT FOR DIFFERENT RENOVATION FREQUENCIES

Year	Initial costs (€)	Registered volume (m³)	Present value of annual revenues from registered water (€)	Present value of cumulative revenues from registered water (€)	NPV of revenues (€)	NPVC of revenues (€)
1	−29	112.80	33.84	33.84	4.84	169.4
2		112.44	32.77	66.61	37.61	667.7
3		112.08	31.73	98.34	69.34	832.5
4		111.72	30.72	129.06	100.06	914.0
5		111.36	29.75	158.81	129.81	962.1
6		111.00	28.81	187.62	158.62	993.6
7		110.64	27.89	215.51	186.51	1015.6
8		110.28	27.01	242.52	213.52	1031.6
9		109.92	26.15	268.67	239.67	1043.7
10		109.56	25.32	293.99	264.99	1053.1
11		109.20	24.52	318.51	289.51	1060.4
12		108.84	23.74	342.25	313.25	1066.2
13		108.48	22.98	365.23	336.23	1070.9
14		108.12	22.25	387.48	358.48	1074.7
15		107.76	21.54	409.03	380.03	1077.7
16		107.40	20.86	429.88	400.88	1080.2
17		107.04	20.19	450.08	421.08	1082.2
18		106.68	19.55	469.63	440.63	1083.9
19		106.32	18.93	488.56	459.56	1085.2
20		105.96	18.33	506.89	477.89	1086.2
21		105.6	17.74	524.63	495.63	1087.0
22		105.24	17.18	541.80	512.80	1087.6
23		104.88	16.63	558.43	529.43	1088.0
24		104.52	16.10	574.53	545.53	1088.3
25		104.16	15.58	590.11	561.11	1088.4
26		103.8	15.09	605.20	576.20	**1088.5**
27		103.44	14.60	619.81	590.81	1088.4
28		103.08	14.14	633.94	604.94	1088.3
29		102.72	13.69	647.63	618.63	1088.1
30		102.36	13.25	660.88	631.88	1087.8

TABLE 7.5. NPVC AND RENOVATION PERIOD FOR THREE METER MODELS

	Model A	Model B	Model C
NPVC (€)	1088.5	1091.9	1095.8
Renovation period (years)	26	28	31

From an economical point of view, the most profitable meter is Model C. However, the optimum renovation period, over 30 years seems to be too long. In practice, such a renovation period would mean that a good number of meters would already be stopped at that time, and this would be the actual time for a replacement.

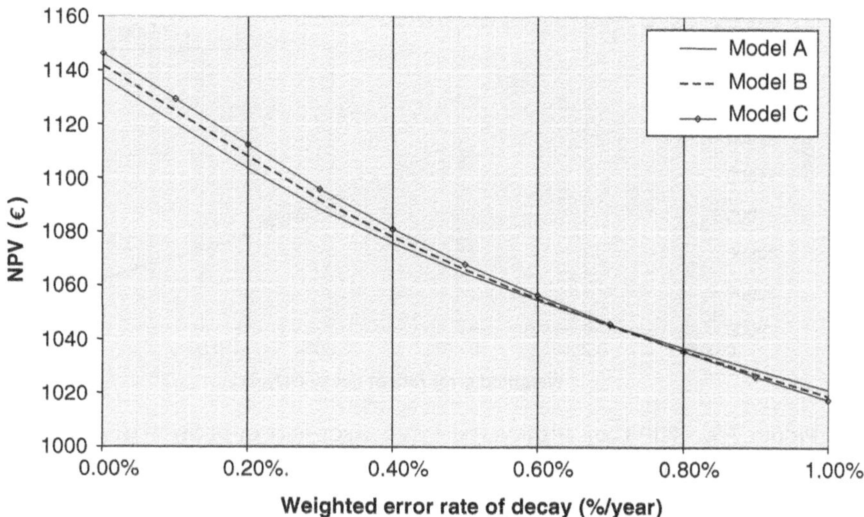

FIGURE 7.8.   NPVC OF THE INVESTMENT VS. ERROR RATE OF DECAY

Furthermore, for such periods, the assumption made of a linear and constant rate of decay for the weighted error will not longer be valid.

If the NPVC of the revenues is now plotted against the decay of the weighted error (Figure 7.8) it can be seen that the differences are smaller than it was before (Figure 7.4). This can be explained by the fact that each meter is optimized separately with an optimum renovation period. In this method, and for a given rate of decay, each meter would remain in use a different number of years that would maximize the NPVC of each investment.

It can also be seen that the differences in the rates of decay of the weighted error are not constant anymore. Depending on the rate of decay, the optimum meter changes. If the expected rate of decay is 0.3%/year, the best meter is Model C. All meters become equal for a rate of decay of 0.7%/year (the NPVC of the investment for all models becomes identical). For higher rates of decay, Model A provides higher benefits than the other meters.

If the discount rate is set to the value of inflation, the differences grow and Model C becomes the best option in all cases (Figure 7.9).

A similar result is obtained when the price of the cubic meter is increased to 1€/m^3. The higher the price, the more profitable Model C becomes (Figure 7.10).

7.4.3.   Estimation of the lifespan

Using Equation (7.5) and simulating multiple scenarios with different initial conditions, it is possible to estimate the lifespan of a meter (the period of time that will

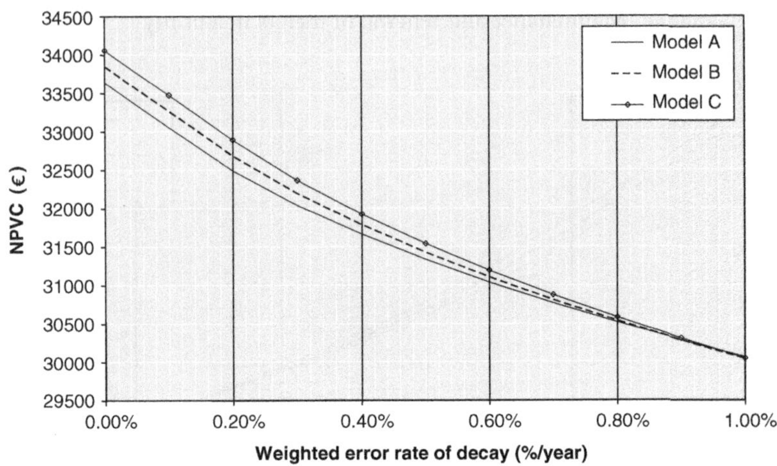

FIGURE 7.9. NPVC OF THE INVESTMENT VS. ERROR RATE OF DECAY WITH A 2.1%
DISCOUNT RATE

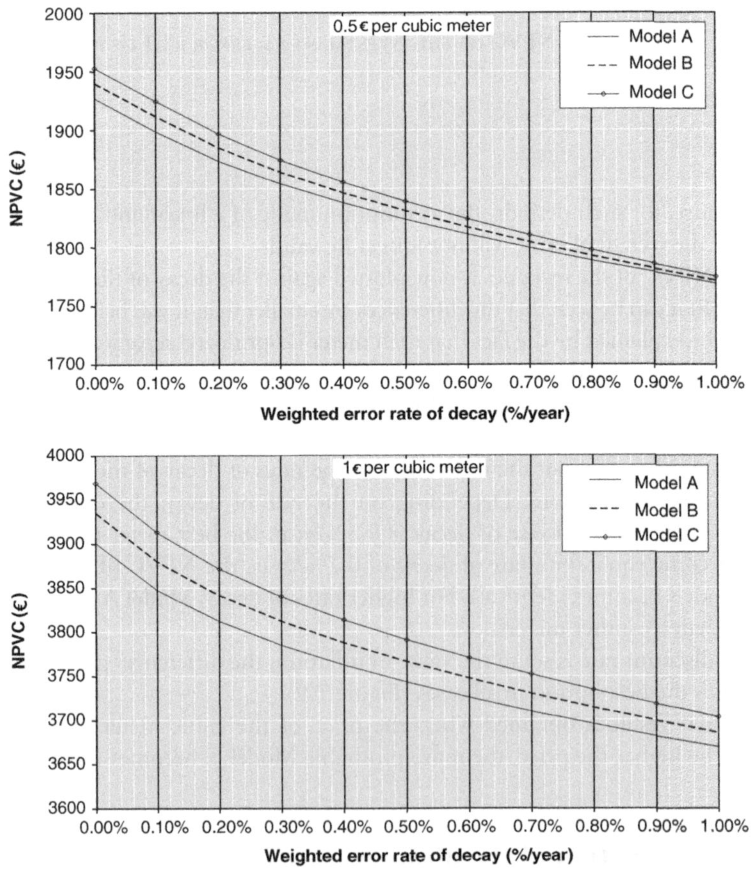

FIGURE 7.10. NPVC OF THE INVESTMENT VS. ERROR RATE OF DECAY FOR DIFFERENT
PRICES OF WATER

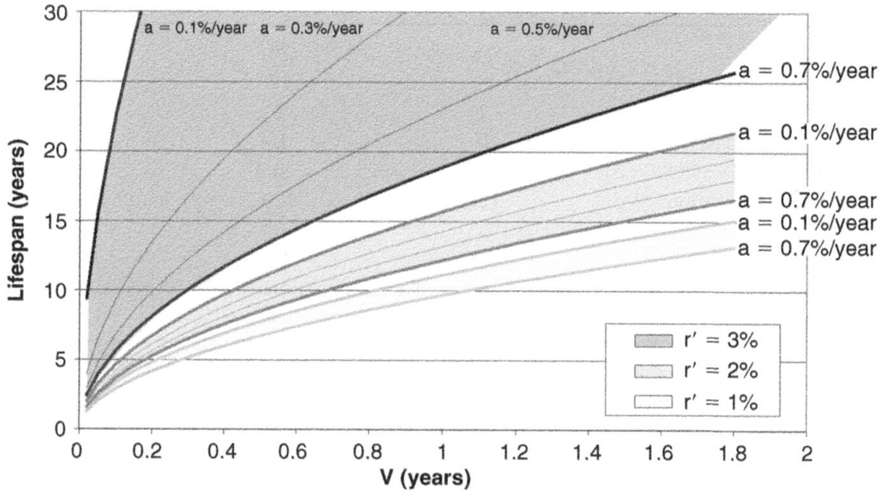

FIGURE 7.11. LIFESPAN OF A METER DEPENDING ON THE INITIAL CONDITIONS

maximize the NPVC of the revenues cash flow). More specifically, the following values have been considered for the different values in the model:

- Rate of decay of the weighted error (a): 0.1, 0.3, 0.5 and 0.7%/year.
- Real discount rate (r'): 1%, 2% and 3%.
- V ratio: 0.05–1.8 years.

Where:

$$V(\text{years}) = \frac{\text{Meter cost } (\text{€}) + \text{Installation costs } (\text{€})}{\text{Consumption } (\text{m}^3/\text{year}) \cdot \text{Water price } (\text{€}/\text{m}^3)} \tag{7.7}$$

Changing these parameters and calculating the NPVC the optimum lifespan of each meter is obtained. The results are summarized in Figure 7.11.

7.5. CASE STUDY

As an example, the characteristic variables of a meter and a utility are shown in Table 7.6.

From Figure 7.12 the lifespan of each model can be calculated from the values of the different variables.

If the decay rate increases from 0.3%/year to 0.5%/year (with all the other variables remaining constant), the lifespan of all three models would be reduced (Figure 7.13).

TABLE 7.6. CHARACTERISTIC VARIABLES FOR THREE METER MODELS

	Model A	Model B	Model C
Retail price (€)	15	20	25
Installation costs (€)	12	12	12
Administrative costs (€)	2	2	2
Metrological class	B	B	C
Error rate of decay (%/year)*	−0.3	−0.3	−0.3
Annual consumption (m^3/year)	120	120	120
Price of water (€/m^3)	0.3	0.3	0.3
Real discount rate (r')	Approximately 3%	Approximately 3%	Approximately 3%
V ratio	0.81	0.94	1.03

* Previously estimated.

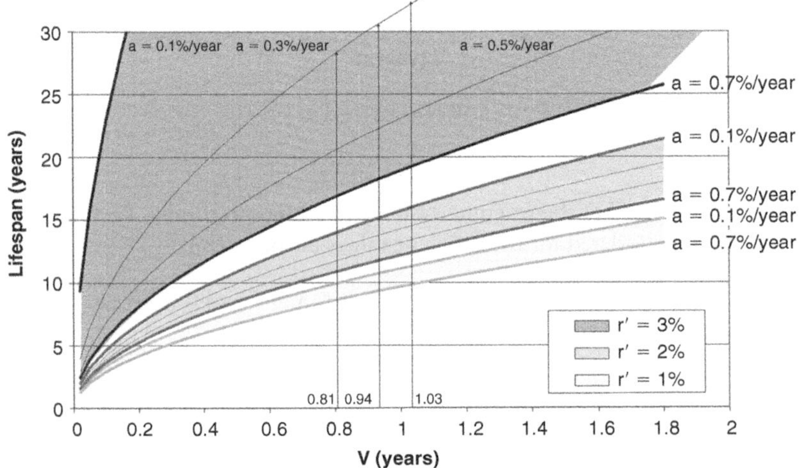

FIGURE 7.12. LIFESPAN OF THE METERS FROM THE CASE STUDY

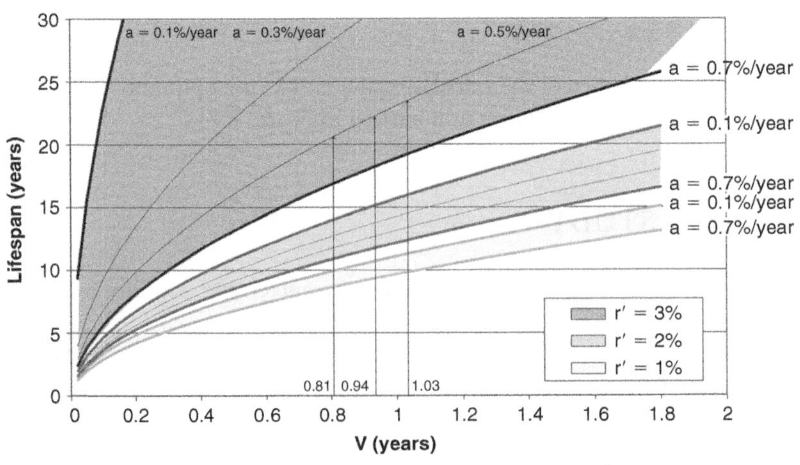

FIGURE 7.13. LIFESPAN OF THE METERS WITH A RATE OF DECAY OF 0.5%/YEAR

7.6. IS IT WORTH GOING FROM CLASS B TO CLASS C?

Another recurrent question that puzzles utility managers is whether to stay with a Class B meter or upgrade to a Class C with the obvious increase in cost. This question is answered by determining the admissible cost increase in changing classes for a certain case.

In order to find a precise answer to this question, Equation (7.5) must be solved for different meter prices and take into account the increase in accuracy from the Class C meter (usually about 2% or 3%). However, in utilities with very adverse consumption patterns, this gap could reach 5%. On the contrary, in utilities where leakage inside the buildings is unlikely this figure could drop to nearly 1%.

Solving Equation (7.5) repeatedly can turn into a tedious process, and a computer software to carry out the process is often advised. However, if such a software is not available, the problem can be simplified by means of Equation (7.6). This formula will determine the admissible extra cost, for a better performance meter, accurately enough $(-1€$ to $+3€)$ for a value of the V ratio above 0.4 (which is usually the case) and a lifespan under 20 years:

$$\Delta\begin{pmatrix}\text{Admissible}\\\text{cost}\end{pmatrix}(€) = \begin{pmatrix}\text{Water}\\\text{consumption}\end{pmatrix}\begin{pmatrix}\dfrac{m^3}{\text{Year}}\end{pmatrix}\cdot\begin{pmatrix}\text{Water}\\\text{price}\end{pmatrix}\begin{pmatrix}\dfrac{€}{m^3}\end{pmatrix}$$
$$\cdot\begin{pmatrix}\text{Accuracy}\\\text{improvement}\end{pmatrix}(\text{p.u.})\cdot\text{Lifespan(years)} \qquad (7.8)$$

In order to use Equation (7.6), the lifespan of Class B meter must first be determined using Figure 7.11. However, the rate of decay is not known beforehand and in an absence of real data a typical value (e.g. 0.3%/year or 0.5%/year) will have to be estimated. The accuracy improvement concerning the meters can be estimated depending on the type of user of the utility (this value is often close to 0.02 p.u. or 2%).

The following example refers to a utility with consumption patterns which do not have excessive volume at low flowrates. From the Class B meter and utility data it is possible to determine the admissible price that a Class C meter would need to have to make it a better option from an economic point of view.

The estimated lifespan for the Class B meter is 14 years (Figure 7.14). Taking into account that the gap in accuracy when using a Class C meter is of about 2%:

$$\Delta\text{Admissible cost }(€) = 140\begin{pmatrix}\dfrac{m^3}{\text{year}}\end{pmatrix}\cdot 0.5\begin{pmatrix}\dfrac{€}{m^3}\end{pmatrix}\cdot 0.02\cdot 14(\text{year}) = 19.6€$$

Maximum cost of the meter = $15€ + 19.6€ = 34.6€$.

Using Equation (7.5), the NPVC of the cash flow revenues for the first option (Class B meter) is obtained: 2108.6€. The NPVC of the scenario using the Class C meter with a lifespan of 19 years would be of 2106.6€.

The exact solution to the problem is shown in Figure 7.15, where Equation (7.5) is solved for several meter prices. Figure 7.15 also shows that the maximum price that should be paid for a Class C meter, which improves the registering error a 2%,

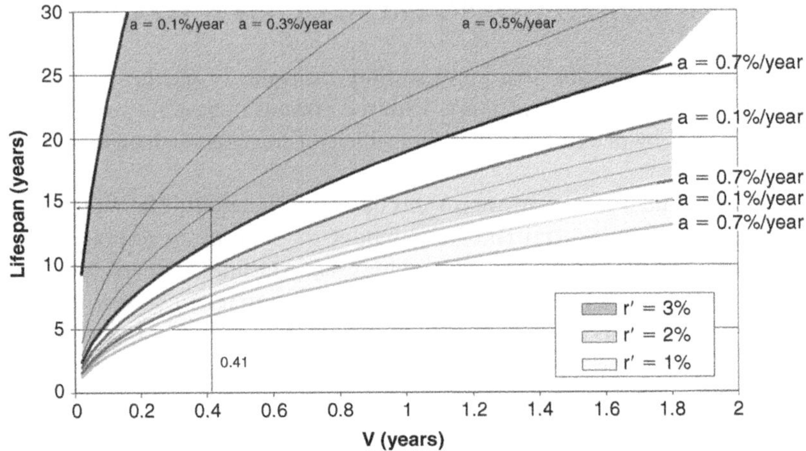

FIGURE 7.14. ESTIMATION OF THE LIFESPAN FOR THE CLASS B METER

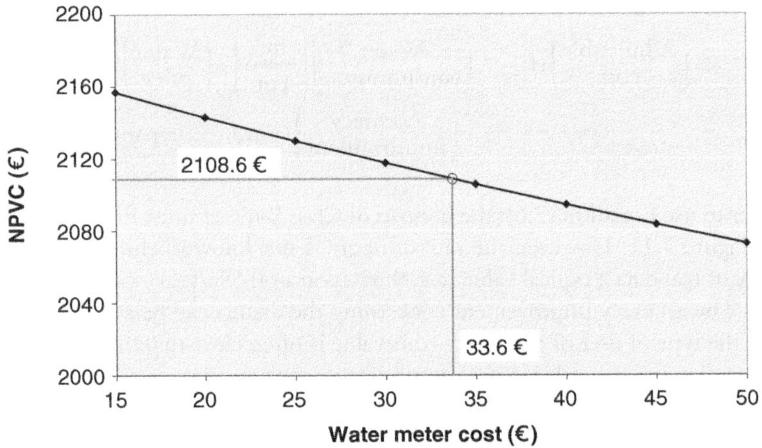

FIGURE 7.15. MAXIMUM ADMISSIBLE PRICE FOR A CLASS C METER (ASSUMING IT IMPROVES METERING ERRORS BY 2%)

to obtain the same NPVC of the Class B meter should be of 33.6€. In this case, using the estimation formula led to an error of +1€.

In any case, it must be taken into account that for this example, it has been assumed that the weighted error rate of decay for both metrological classes, as well as the installation and administrative costs are the same.

7.7. USING THE COMMERCIAL INFORMATION SYSTEM TO ANALYSE THE BEHAVIOUR OF METERS

The commercial information system of a utility (database storing all data related to billing) may be turned into a powerful tool that can lead to valuable conclusions on the

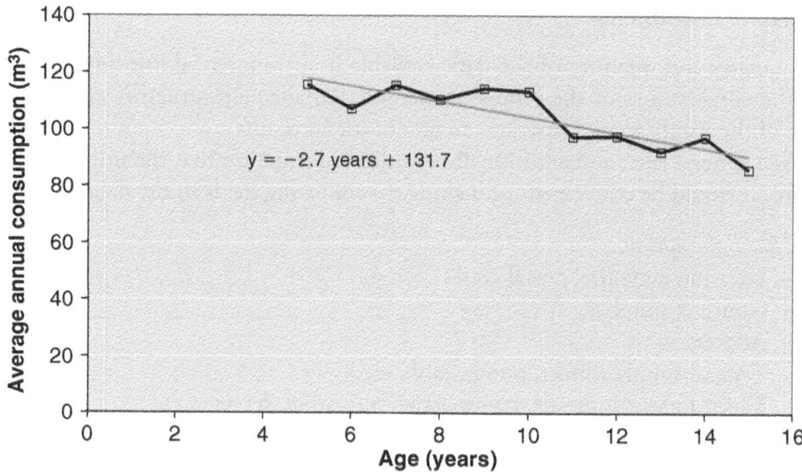

FIGURE 7.16. EVOLUTION OF THE AVERAGE DAILY VOLUMES BILLED AS A FUNCTION OF AGE

behaviour of the meters, including estimations of their rate of decay and the probability of being stopped.

As has already been defined, the weighted error of a meter is the percentage of the consumed volume that it is served but not metered. This value can be estimated from the error curve of the meter and the user's consumption pattern (Chapter 6).

The method shown here is based on the fact that the metering error of a meter is reflected in the average volume registered per unit of time. In other words, the least accurate meters, in average, will register less volume than the more accurate meters. However, in order to obtain significant results, the number of meters studied must be large enough depending on the dispersion of user types.

Consequently, in principle, analyzing the billing data for a certain type of meter, depending on the age and the type of user, it is possible to obtain an estimation of the evolution of its error. Figure 7.16 shows an example of the type of information that results from applying this procedure, with an evolution of the bills resulting from a meter as it gets older.

This method does not allow to determine the global error directly, but it can predict the rate of decay in the accuracy of the meter as a function of age and provides a tool to compare several models.

Using the commercial information system of a utility to make this sort of predictions has several obvious advantages. These systems are organized in databases that are updated with reliable information (the utility revenues depend on them). Once the analysis procedures have been identified, it is possible to develop custom made software packages that can periodically analyse all meters in a utility. Information can be generated almost on real time, and reactions to possible deficiencies in a meter model can be quickly corrected with the consequent savings.

In practice, using the commercial information system will not completely eliminate the field samples, but it represents an ideal complementary tool to analyse and manage all meters in a utility.

7.7.1. Databases

The quality and quantity of the data available in a commercial information system are the cornerstones of the proposed method. An adequate structure and maintenance of the database are important to obtain useful results.

As a general rule, and additionally to other information that the utility may find useful, it would be convenient to include the following fields in the database:

- *User information*
  - Location code (i.e. postal code)
  - Contract number
  - Address
  - Type of supply (direct, pump, tank, etc.)
  - Type of user (domestic, commercial, industrial, services, etc.)
  - Observations
- *Meter information*
  - Nominal flowrate
  - Metrological class
  - Diameter of meter connections
  - Length
  - Full serial number
  - Manufacturing year
  - Manufacturer
  - Model
  - Type of meter (single jet, multiple jet, Woltmann, etc.)
  - Last revision
  - Installation date
  - Place of installation (indoor, outdoor, etc.)
  - Type of installation (ground, wall, roof, etc.)
  - Observations
- *Reading information*
  - Reading date
  - Registered volume
  - Reader
  - Route
  - Type of reading (real, estimated, user, etc.)
  - Observations

Since the integrity of the data is essential, the utility must develop the systems and protocols necessary to guarantee the reliability and accuracy of the data, especially regarding to the registered volumes, installation dates, serial numbers and models. These variables will be the ones used to query the database.

7.7.2. Using the commercial information system

The method is mainly based on analysing the evolution of the average yearly registered volume – any other period can be used – as a function of the age of the meter (time installed) or the accumulated volume. Additionally, using the water consumption information, it will be possible to detect meters too large or small for the user.

Querying the database, the meters should be grouped by models and age classes, and an average value of the consumptions of those users can be calculated. In order for the procedure to be successful, certain conditions must apply:

- The meters under study must be homogeneously installed in the utility. This means that variables such as the characteristics of the pipes, pressure, quality of water, etc. should be similar throughout the sample.
- Water consumption of the users of different models should be the same. If a meter model is preferable installed in a certain type of user the results will be biased by the differences in user's behaviour.
- The time reference used to compare the different consumptions should be the same.
- The consumption should not be affected by seasonal variables, and if so, those variables should affect all models in the same way.

The database should then be queried according to several factors. Some examples are provided here. It is important to point out that certain interpretation mechanisms should be applied and that these results could be biased one way or the other depending on the query.

Average yearly volume as a function of age

This query groups meters by model and age range, obtaining for instance the average yearly consumption (or monthly). Following this procedure it is possible to determine the evolution of the weighted metering error as a function of the meter age. The ratio between the constant term and the slope of the linear adjustment of the curve is an indicator of the rate of decay. In the real case shown in Figure 7.16, this figure is approximately -2.1% ($-2.7/131$). This value is higher than the one usually found in meters. However, it should be taken into account that this figure would also consider stopped meters.

These results however can be biased:

- When grouping by age ranges, the compared meters belong to different manufacturing dates. The design improvements may have the effect of an improved rate of decay for the newer models.
- The location of meters may not be homogeneous for all age groups. If a certain group is more present in a sensitive area, they could show higher rates of decay and produce graphs without a clear trend.
- It is difficult to guarantee that the characteristics of users are independent of the meters' age. For instance, the average household size in many countries has decreased due to increasing costs.

In order to avoid many of these inconveniences, the same sort of query may be done individually, and later find the average of the registered volumes' evolutions. This procedure eliminates the influence of the characteristics of users, but requires several billing periods to obtain results. Consequently, it can only be applied when the meters are being tracked from the moment of their installation. The procedure also requires

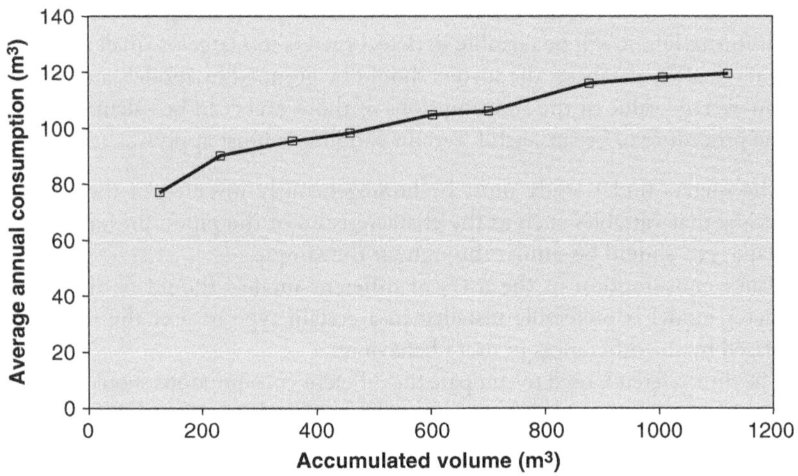

FIGURE 7.17.   EVOLUTION OF AVERAGE DAILY BILLED VOLUMES AS A
FUNCTION OF ACCUMULATED VOLUME

to correct the possible variations in the demand due to climatic, social and economic changes.

Average daily volume as a function of accumulated volume

Determining meter wear from the accumulated volume instead of time usually provides better correlations with the average registered volume. However, this sort of query does not allow to determine the rate of decay and is only useful to compare models. The good news is that it eliminates the bias created by the users' characteristics and the age of meters.

This query will always produce a positive slope. A reasonable expectation would have been to find negative slopes due to the increased error with age. The explanation for this apparently anomalous behaviour is that the meters with a higher accumulated volume are usually those of the larger users. The meter of a user with a yearly consumption of $120\,m^3$ will hardly ever reach the $3000\,m^3$ of accumulated volume. Although older meters will have a higher accumulated volume, usually the size of the user has a greater impact on this parameter.

The main limitations and biases on this slope are:

- When comparing the behaviour of several meters, slopes are only comparable when the models were installed simultaneously. It is consequently not possible to compare older models with newer ones, even though the comparison is made through the accumulated volume.
- Since the rate of decay is not obtained, it is not possible to obtain an estimation of the metering error.

As in the previous case, it is interesting to follow the evolution of these graphs with time. As a general rule, and with the exception of great variations in the demand, the slope of the regression curve should remain constant.

Large customers

Determining the optimum size of meters for large customers is a key factor to reduce the overall metering error. Several techniques to determine the nominal flowrate of a meter have already been presented. The most frequent one due to its simplicity uses the average monthly consumption to select the most appropriate meter. Several manufacturers offer selection tables in which ranges of typical monthly consumptions are presented for each nominal flowrate. These values can be used as a reference to identify meters which are either too small or too large for a specific user.

Users are in this case studied individually after being selected through a query. The method can be complemented by setting certain trigger rules that may give warning in case of sudden changes in consumption from 1 month to the next one (and for instance give warning of possible leaks to the users). The analysis of monthly data will also allow to identify possible anomalies in the meter.

Generally speaking, large customers' meters should be read more than once a month, and with a frequency that should increase with the user's size. As a matter of fact, using electronic meters and data transmission equipment usually pays off for these users as the investment is usually recovered in a short time. Additionally the use of statistical analysis tools may significantly shorten the time needed to detect anomalies, reducing the chance of human error in the interpretation of data (Kurokawa and Bornia, 2002; Palau et al., 2003).

Example

The following example shows how useful can the commercial information system. As a matter of fact, the information it can provide matches the results obtained through more traditional methods including testing samples and measuring consumption patterns.

In this case study, the commercial information system of a utility was queried in order to assess the behaviour of two types of meters in a city. More specifically, the objective was to determine the differences between multiple jet meters and oscillating piston meters.

Figure 7.18 shows the yearly average consumption registered in a billing period vs. the volume accumulated by the meters. In both cases the resulting slope is positive. Additionally, Figure 7.19 showed that the average yearly consumption (the ratio between the accumulated volume and the time lapsed from the meter's installation) was higher for those meters with a higher accumulated volume. This ratio was found to be similar for both types of meters, showing that the characteristics of users equipped with both technologies were similar.

From Figure 7.18 it can be deduced that the yearly registered volume is higher in multiple jet meters than in volumetric meters. As meter get older (and they accumulate more volume) these differences increase.

The commercial information system analysis was complemented by a series of laboratory tests aimed to determine the meters' error and the consumption pattern of the users. A total of 248 oscillating piston meters and 349 multiple jet meters were tested using different test flowrates. The results showed that the error of volumetric meters was always negative, while the error of multiple jet meters at medium and high flowrates tended towards positive values (as they registered more water than it was actually consumed) (Figures 7.20 and 7.21).

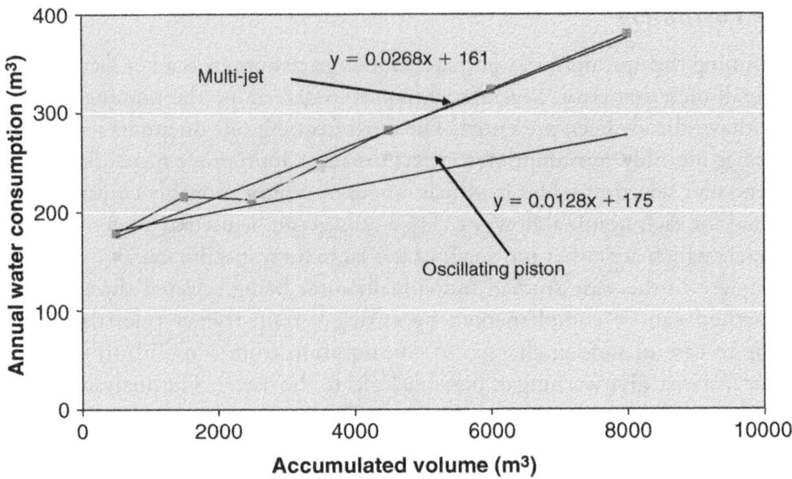

FIGURE 7.18.   YEARLY REGISTERED VOLUME VS. ACCUMULATED VOLUME

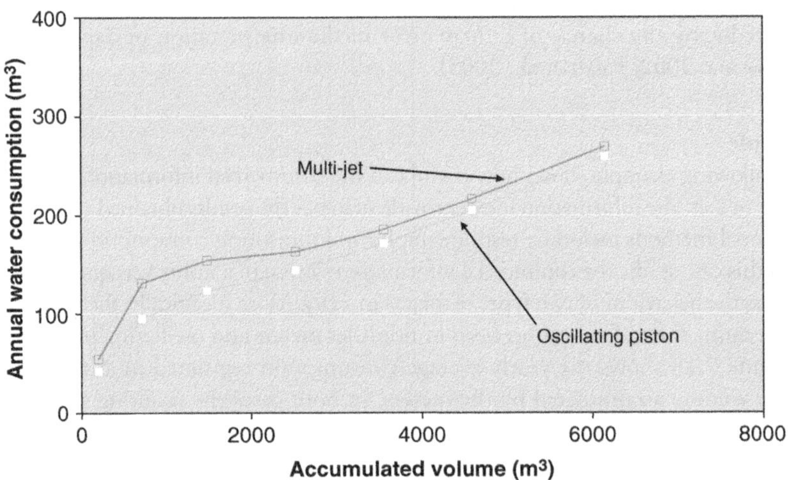

FIGURE 7.19.   AVERAGE CONSUMPTION VS. ACCUMULATED VOLUME

By weighing these curves with the characteristic demand pattern of the population obtained by measuring consumption in 305 households, it was possible to obtain the evolution of the global metering error. The results are shown in Figure 7.22. The volumetric meters' error decreases (increases towards negative values) significantly with the accumulated volume, while the multiple jet meters maintain their error quite constant through time.

Table 7.7 compares the results obtained by analysing the data of the commercial information system (Figure 7.18) and using traditional methods (Figure 7.22). In order to determine the billing differences between both types of meters, a linear fit of the data was performed (other types of fitting could be used). This type of analysis highlights even more the higher performance of velocity meters.

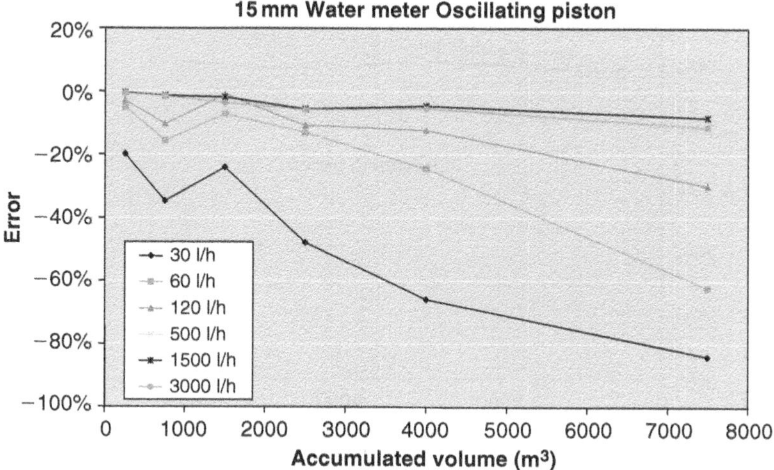

FIGURE 7.20.   EVOLUTION OF THE METERING ERROR AT DIFFERENT FLOWRATES
(15 MM WATER METER, OSCILLATING PISTON)

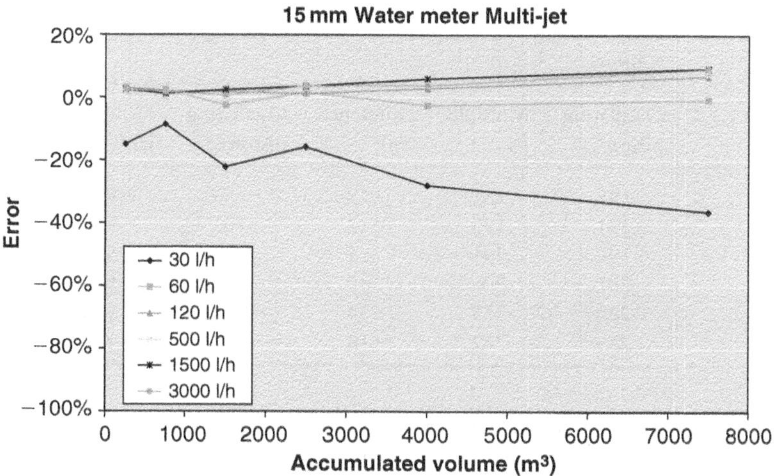

FIGURE 7.21.   EVOLUTION OF THE METERING ERROR AT DIFFERENT FLOWRATES
(15 MM WATER METER, MULTIPLE JET)

As an example, using the linear regressions carried out on the results from the queries to the commercial information system, it is known that an oscillating piston meter that has an accumulated volume of $1500\,m^3$ will only bill $194\,m^3$ vs. the $201\,m^3$ of a multiple jet meter. This difference is approximately 4% less in volume and, with linear tariffs, 4% less in revenue.

From the evolution of the global error with time it can be deduced that the average error of a meter with $1500\,m^3$ of accumulated volume is approximately -11.3% in the case of oscillating piston models and -5.2% in multiple jet ones. These differences are not too significant, although they are higher for values of the accumulated

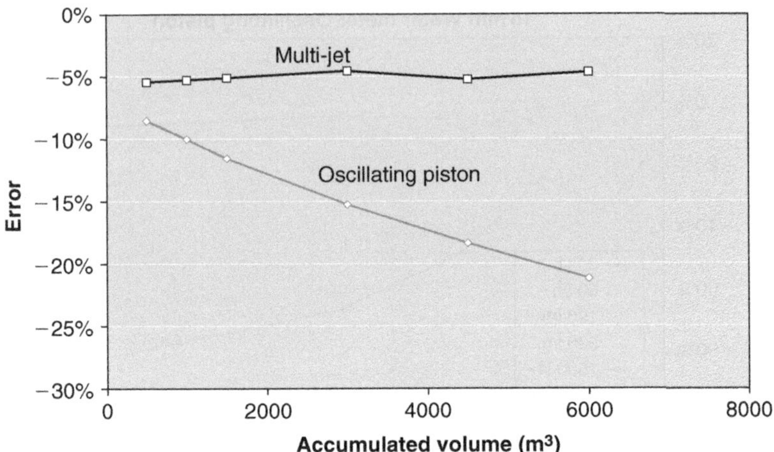

FIGURE 7.22. EVOLUTION OF THE GLOBAL ERROR WITH THE ACCUMULATED VOLUME

TABLE 7.7. COMMERCIAL INFORMATION SYSTEM ANALYSIS VS. TRADITIONAL METHODS

Accumulated volume (m³)	Commercial information system analysis (Figure 7.18) registered volume			Traditional methods (Figure 7.22) metering error		
	Oscillating piston	Multiple jet	Difference (%)	Oscillating piston (%)	Multiple jet (%)	Difference (%)
750	185	181	2	−9.6	−5.3	−4
1250	191	195	−2	−10.7	−5.2	−5
1500	194	201	−4	−11.3	−5.2	−6
2500	207	228	−9	−13.6	−5.1	−9
3500	220	255	−14	−15.9	−5.0	−12
4500	233	282	−18	−18.2	−4.9	−14

volume lower than $750\,\text{m}^3$ or higher than $4500\,\text{m}^3$, always underlining the superior performance of multiple jet models.

The quality of the results mainly depends on the accuracy of the data present in the system. In this case, the database had not been adequately maintained for a long time, and it was difficult to filter out wrong data and possible biasing elements. On the other hand, the estimation of errors by traditional methods is not free of uncertainties, mainly due to the small sample sizes. Some of the discrepancies between both methods could also be due to this fact.

The analysis of the commercial information system is able to determine this factor as long as there is accurate and updated data on the installed meters and their billing values. Most of the time, the lack of adequate maintenance in the users' database does not allow a proper analysis and no useful results can be obtained. However, and even in those cases, it is still possible to use the method (being as cautious as it is necessary) to compare the performance of several types of meters.

8

Meter testing

8.1. INTRODUCTION

The metrological control of water meters, both new and used, is one of the cornerstones of meter management. Almost all existing methods to analyse the global error of all meters in a utility require information resulting from testing individual meter errors of a defined sample.

Consequently, it is vital to approach error testing systematically and with a reduced uncertainty. After all, quite often the results of such tests will be used afterwards without any further questioning on their validity. It is therefore a sensitive process that could lead to the wrong conclusions and generate considerable economic losses.

This chapter will present the basic aspects of the procedures and the testing equipment needed to provide reliable and accurate results.

8.2. TYPES OF TEST BENCHES

The first step in this process is the availability of a testing laboratory, either owned by the utility or by a third party. The objectives of the tests should be clear well in advance, and so should be the device population subject to the tests. These details will condition the requirements of accuracy for the test bench, as well as its capacity. Quite obviously, the economical decision of whether a utility should own a laboratory or not will depend on the number of annual tests carried out. After all, the maintenance of such a laboratory is a cost intensive activity and will only be profitable when the number of tests is high and the use of the laboratory constant.

The determination of the meter's error at a certain flowrate is done by comparison of the registered volume of the tested meter and a reference volume measured with a lower uncertainty. The reference volume can be determined in a number of ways (for instance a high accuracy flow meter, a tank of a known volume or the weight of the water that has circulated through the meter in a certain amount of time). These last two methods are obviously the most accurate ones, for in the first case the uncertainty is limited by the metrological performance of the flow meter used.

8.3. TEST BENCH LAYOUT

Basically speaking, a test bench (Figure 8.1) consists of an suction tank, a group of pumps, a hydro-pneumatic tank (optional), one or several filters, control and operation valves, flow meter(s) used as reference instruments for tests or simply to set the value of the circulating flowrate and a section to set up the meters to be tested. Additionally, pressure and temperature transducers can be installed. At the end of the testing line a weighting instrument or a volumetric tank can be placed.

The suction tank must be large enough to avoid cavitation in the pump when the largest diameter section (the one requiring a largest amount of water) is full. The volume

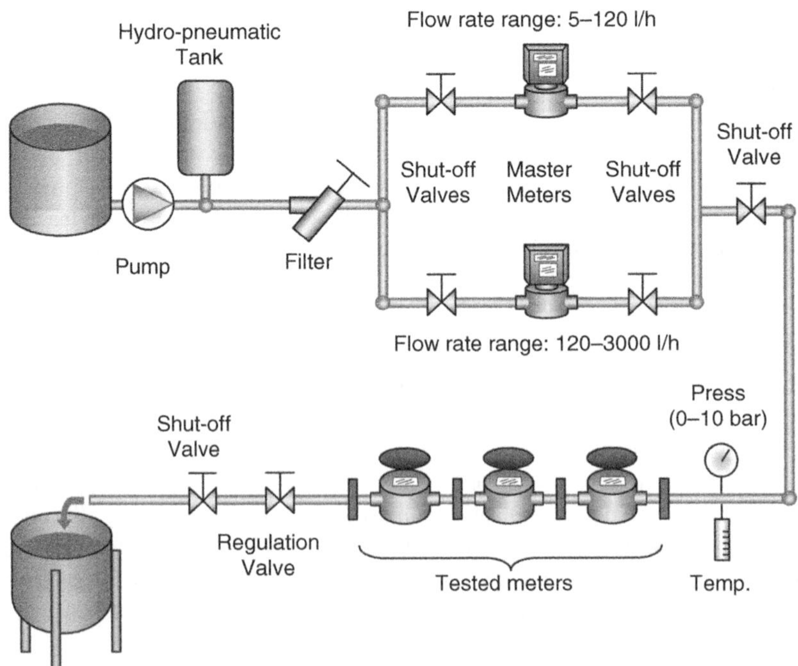

FIGURE 8.1. TEST BENCH SCHEMATICS

of the tank should enable the use of the pumps at maximum flowrate during the tests. The tests may last all day long and a small tank could lead to an increase in water temperature above the required test levels. This problem can also be solved through a greater thermal dissipation capacity of the whole system, including tank and pipes.

The installed pumps may function at a constant velocity, although it is recommended to install, at least, a variable speed pump which will allow setting the test flowrate to the desired value and reducing the energy consumption of the bench. The velocity of the pump can be set manually, depending on the expected testing conditions or automatically, as a function of a pressure or flowrate signal. This last procedure, which in principle might appear to be better, could lead to unstable testing conditions subject to the regulation parameters of the pump. Depending on the desired operating flowrate range of the test bench it may be necessary to install several pumps in parallel.

The head delivered by the pump should be enough to overcome the pressure losses created by the meters and the bench itself. Generally speaking, meter losses at maximum flowrate can be estimated in almost 1 bar. An estimation of the minimum pump head required at the maximum test flowrate can be made by:

$$H_{\mathrm{b}} \text{ (in bar)} = \text{(number of meters tested)} \cdot 1.2 \tag{8.1}$$

Following this recommendation, if a test bench has a capacity of five meters of nominal flowrate of $1.5\,\mathrm{m^3/h}$ the minimum head to be provided by the pump, to test the meters at maximum flow ($3\,\mathrm{m^3/h}$), would be approximately $7.2\,\mathrm{bar}$. If, additionally, the same bench must have the capacity to test three meters with a nominal flowrate of $10\,\mathrm{m^3/h}$, the pump should provide $3.6\,\mathrm{bar}$ at $20\,\mathrm{m^3/h}$.

Downstream from the pump it is useful to install a *hydro-pneumatic tank* to reduce pump vibrations, which could ruin the test results especially at low flowrates. If physical space and budget allow it, a large hydro-pneumatic tank may also feed the tests. The tank provides a fairly constant hydraulic head with the pumps stopped. This solution results in vibration-free tests. In such case, the only precaution that must be taken is to maintain the pressure levels at the tank in order to keep the test flowrates within the standardized limits. A pressure reducing valve downstream the hydro-pneumatic tank will also help maintaining a constant flowrate throughout the test.

Upstream from the tested meters, a filter must be installed to prevent damage to the devices from particles that may have accumulated in the aspiration tank. The presence of a filter becomes crucial when the control devices are electromagnetic or volumetric flow meters of a reduced diameter (2 or 3 mm).

Depending on the role of the control instruments (whether they are just used to control the flowrate or as the actual reference for the test) their accuracy will be very different. For flow control purposes the required accuracy is only 2%, but should they be used as the reference for the test, the maximum allowed error in the operating range cannot exceed 0.2% (one-tenth of the maximum permissible error of a meter). It is important to point out that this error cannot be exceeded at any flowrate. Consequently, the selection of the device becomes a very sensitive issue and the operating ranges of each device need to be carefully selected. The usual choices are oscillating piston meters (up to 40 mm) or electromagnetic flow meters.

FIGURE 8.2. WATER METER TESTING BENCH

Control devices can be installed upstream or downstream from the meters. Installation conditions should satisfy the manufacturer's requirements, especially those concerning the length of straight pipe upstream from the meters.

Usually test benches allow setting up several meters in series, and incorporate the necessary fittings to accommodate several diameters and lengths. Figure 8.2 shows the testing section of a bench with five 13/15 mm meters.

The testing section must have the possibility to be hydraulically isolated from the rest of the facility with the adequate valves. Downstream of the testing section, the flowrate is controlled by means of regulating valves. These valves should be carefully selected and be able to set and maintain stable low flowrates. The requirements at medium and high flowrates are much less strict.

Following specific characteristics of the different types of test benches are commented.

8.3.1. Using a flow or master meter as reference

This type of bench is the cheapest to construct, and is therefore a good option if economic resources are limited. However, and depending on the number of flow meters that are necessary (which is related to the operational flowrate range of the bench) its cost may not be far from the other two options. This alternative results in higher uncertainties associated to the tests, and it is unlikely that they can be kept under the limits set by standards or national regulations (lower than 0.2%). For this reason, this type of bench must only be used to test meters retired from the field, and never as a control tool for new meters.

TABLE 8.1. ELECTROMAGNETIC METERS NEEDED IN A BENCH FOR FLOWRATES BETWEEN
10 AND 40,000 L/H, WITH AN UNCERTAINTY OF 0.5%

DN (mm)	Minimum velocity (m/s)	Maximum velocity (m/s)	Minimum flowrate (l/h)	Maximum flowrate (l/h)
2	0.3	10	3	113
15	0.3	10	191	6362
40	0.3	10	1357	45239

The reference devices must be periodically calibrated. Usually, several flow meters are needed in order to cover the flow meter range designed for the bench with the required accuracy. As an example, consider a bench designed to verify meters at flowrates ranging from 10 to 40,000 l/h. In this case, and taking into account that the admissible velocity range for an electromagnetic flow meter is from 0.3 to 10 m/s, the devices listed in Table 8.1 would be necessary. Additionally, all these meters should provide the requested accuracy in all the measuring range.

Another issue to be taken into account is the testing procedure. For instance, the control devices must be able to correctly integrate the flowrate during the duration of the test. In order to guarantee this, the integration interval of the flowrate signal should be adequate and so the equipment response to changes in flowrate values. Additionally, the reading of the totalized volumes should be easy, updated and with enough resolution to ensure the required uncertainty. Finally the reference instrument should be equipped with a reset button.

8.3.2. Collection in a volumetric tank

In this configuration, the control element used as a reference is a tank of known volume. Usually, each bench will be equipped with one or two tanks, allowing the tests to be performed on a wide range of volumes. The parameters that condition the volume to be used in the tanks are mainly two: the resolution of the tested meter and the time required for the test.

The first parameter has to do with the uncertainty associated to the reading of the meter and depends on its characteristics. For instance, if 10 l of water are used for a test and the reading of the meter can be assessed with a resolution of 0.2 l, the uncertainty could be assessed as the ratio between half the volume associated to the resolution (0.1 l) and the total volume measured in the tank. In this case, the uncertainty would be 1% too far from the 0.2%, which is the value usually required in tests.

The second parameter that should be taken into account when selecting the volume of the tank is the time required to fill it. The biggest disadvantage of this configuration is that the tests have to be carried out by comparing a fixed known volume (the tank's volume). As a result, the time required for each test is obtained by dividing this volume by the flowrate required for the test. For instance if a 1.5 m^3/h Class C meter is tested to its minimum flowrate (15 l/h), by filling a 10 l tank, the time consumed is 40 minutes. Reducing this time would mean using a smaller tank, but the uncertainty associated to the test, originated in the meter reading process could then become too high.

The test bench tanks are usually made of stainless steel in order to avoid changes in volume related to thermal dilatation of the deformation of tank walls. Figure 8.3 shows different examples of tank shapes.

The uncertainty related to the capacity of the tank is strongly related to the resolution in the reading of the level. Normally, achieving a resolution better than ± 3 mm (and hence reducing such value) is complicated. For this reason, and in order to obtain an uncertainty of 0.1%, if the tank is completely cylindrical its height should be over 1500 mm. Figure 8.3(a) is currently becoming less popular for the difficulties in reading the level at certain distance from the floor.

Nowadays, the tank designs usually follow Figure 8.3(b) and (c) shapes. In these cases, smaller tanks (less height) reduce the uncertainty to minimum values, determined by the diameter of the narrower part of the tank. If the diameter of this section is 100 mm and the resolution in the reading is 5 mm, the uncertainty in the assessment of the volume is approximately 0.01%, reducing the necessary height of the tank considerably.

Option (c) allows the combining of two different reference volumes in a single tank. This is especially useful when testing meters at medium and high flowrates, when the critical factor is the total test volume and not the reading resolution of the meter. For instance, consider a 1.5 m^3/h tested at 750 l/h. These meters usually have a reading resolution of 0.2 l, which is the minimum resolution requested by the ISO 4064:1993 Standard. If the test consists in filling a 200 l tank, the required time for the test is 16 minutes. The uncertainty related to the reading of the meter would be 0.05%. However, if the tank volume is 100 l, the time is reduced by half and the uncertainty would rise to 0.1%, which is a perfectly assumable value.

The end of the test is controlled by a level switch installed in the tank that may either close the valve situated downstream from the testing section or divert the water flow to a drain.

The main advantage of this configuration over the other configurations presented here is that the reference element (the tank) hardly needs any sort of calibration. The

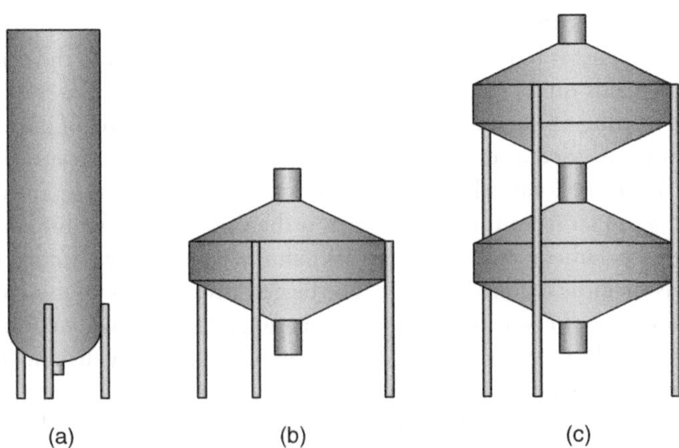

(a)   (b)   (c)

FIGURE 8.3.   VOLUMETRIC TANK SHAPES

maintenance of the tank is usually limited to ensuring that no foreign objects enter the tank and that the dimensional characteristics are not altered.

This configuration is recommended in testing the error curve in small and medium diameter meters. Once the required testing volumes reach certain values, it becomes much cheaper to incorporate weighting elements than a calibrated volumetric tank.

8.3.3. Weighting of the test volume

Another popular option for the configuration of a test bench is to weigh the test volume and use that value as a reference for calibration. The fact that the reference value is in this case a weight and not a volume implies that it needs to be transformed before the comparison with the meters' reading can be made. In the case of water, density is quite constant, although it does change slightly with temperature (Table 8.2). In most applications the density of tap water can be considered to be equal to the density of pure water.

Additionally, it has to be taken into account that the scale "zero" was in fact set with the tank full of air. As a consequence, the real weight of the water filling the tank will be higher than the value shown by the scale (and the weight of air will have to be added). This can also be viewed as a buoyancy effect, where the water displaces the volume of air and consequently experiments an upward force of value equal to the weight of the air.

In practice, both problems are solved by means of a correcting coefficient, K_{air} (Table 8.3), which takes both issues (density and buoyancy) into account. The volume of water filling the tank will then be calculated by multiplying the weight of water by this coefficient which is a function of temperature. For instance, if the reading of the scale is 1000 kg, and the temperature is 21°C, the volume of water filling the tank will be 1003.21 l.

TABLE 8.2.   DENSITY OF PURE WATER AS A FUNCTION OF TEMPERATURE

Temperature (°C)	Density (kg/m^3)	Temperature (°C)	Density (kg/m^3)
0	999.841	16	998.943
1	999.900	17	998.774
2	999.941	18	998.595
3	999.965	19	998.405
4	999.973	20	998.203
5	999.965	21	997.992
6	999.941	22	997.770
7	999.902	23	997.538
8	999.849	24	997.296
9	999.781	25	997.044
10	999.700	26	996.783
11	999.605	27	996.512
12	999.498	28	996.232
13	999.377	29	995.944
14	999.244	30	995.646
15	999.099	31	995.341

TABLE **8.3.**   CORRECTION COEFFICIENT TO OBTAIN THE VOLUME OF WATER
FROM ITS WEIGHT

Temperature (°C)	K_{air} (m³/kg)	Temperature (°C)	K_{air} (m³/kg)
0	1.00145	16	1.00228
1	1.00139	17	1.00245
2	1.00134	18	1.00262
3	1.00131	19	1.00281
4	1.00130	20	1.00301
5	1.00130	21	1.00321
6	1.00132	22	1.00343
7	1.00136	23	1.00366
8	1.00141	24	1.00390
9	1.00147	25	1.00415
10	1.00155	26	1.00441
11	1.00164	27	1.00468
12	1.00174	28	1.00496
13	1.00186	29	1.00525
14	1.00199	30	1.00554
15	1.00213	31	1.00585

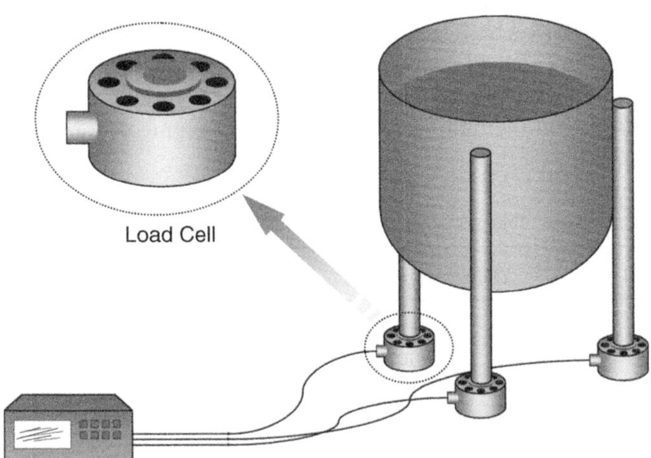

FIGURE **8.4.**   LOAD CELLS-BASED CONFIGURATION

There are basically two options when selecting the instrumentation needed for weighing the volume of water. The first one is based in load cells connected to a module that will add their signals and provide a total output (Figure 8.4).

The second one consists of a scale (Figure 8.5) with a platform, which directly provides the final weight. This second option is more convenient for values up to 1000 or 2000 kg. However, higher mass values will require the first option.

In this configuration, the test bench uncertainty depends mainly on the accuracy of the scale or the load cells. The accuracy in these instruments is usually provided as a

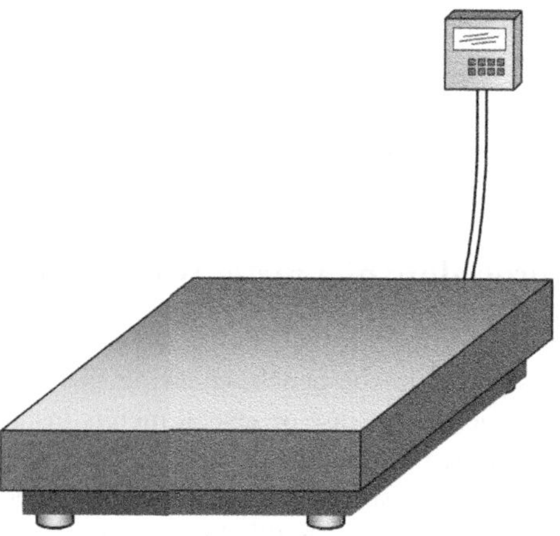

FIGURE 8.5. SCALE-BASED CONFIGURATION

percentage figure of the maximum weight. As a consequence, the relative error associated increases with smaller weights. For instance, if the weight of an object is 10% of the maximum allowable by the scale, the relative error will be 10 times higher than the nominal accuracy of the scale. In any case, uncertainties of up to 0.017% of the maximum value are quite common. In practice this means that weighing objects 5 times lighter than the maximum weighing value can be achieved with relative errors lower than 0.1%.

The high accuracy of the scales allows certain flexibility in the total volume to be used in the tests. This allows reducing of the total time required for the tests, maintaining the uncertainty within reasonable limits.

However, this configuration has the disadvantage of the need of periodic calibrations of the scale or load cells. This implies having a set of calibrated masses of at least 25% of the maximum scale weight. The ISO 9368-1 Standard "Measurement of liquid flow in closed conduits by the weighing method; procedures for checking installations; part 1: static weighing systems" is certainly helpful.

The weighing method is indicated when the total tested volumes are high. In these cases this option turns out to be cheaper than building a large calibrated tank. In almost all facilities devoted to the calibration of large meters this is the method of choice.

8.4. TESTING PROCEDURES

There are two basic ways to determine the error curve of a meter. In the first option, the meter starts from a resting position. No water circulates through the pipe until the valve is open, and the meter starts measuring the volume when the valve is closed again. The registered volume is read before the test begins and after it has finished. Since the meter is at rest, the reading can be done manually without the aid of electronic methods.

A second testing procedure consists in adjusting the circulating flowrate to the desired value while adjusting the control devices. Once everything is ready the flowrate is diverted to the reference device that begins to totalize the circulated volume. When the test is over the flow is again diverted to a drain. In this case, since the reading of the meter must be done while the water is circulating through the pipe, the reading must necessarily be made by means of electronic instrumentation.

8.4.1. Testing procedure of a meter with readings taken with the meter at rest (EN 14154-3)

1. The tested meter is placed in the bench, checking that the flowing direction of water is the correct one. In the case of several meters installed in series, the distance between them must be enough to guarantee that the velocity profile reaching each meter is not being distorted by the upstream meter.

   The testing section is progressively filled with water. The valve situated downstream from the testing section may be open, although it is recommendable to close it. The filling procedure should be undertaken very slowly to avoid changes in the state of the meters due to factors such as limescale build-up, etc. Additionally, filling the testing section rapidly could damage the moving parts of the tested devices.

2. The testing section should be checked for losses and it should be ensured that the temperature and the pressure of water are the adequate ones for the test.

3. The main valve is opened allowing the flow to circulate. The objective is to eliminate the largest possible amount of air that may have been trapped in the meters and the testing section. In this operation, the nominal flowrate should not be surpassed, and in the case of meters retired from the field, the "normal" operating flowrates should be the maximum value to observe. The purge operation must last at least 2 or 3 minutes.

   However, if the test bench is equipped with a vacuum generating system, the test section must be isolated, vacuum should be generated inside and finally allow a gradual filling of the section with water. This is the only fully effective procedure to guarantee that all the air trapped inside the meters is released. As a matter of fact, the purge operation by circulation of water can leave a certain quantity of trapped air in meters with dry or wet totalizers, where some gears are in contact with the fluid, distorting the real behaviour of the meter, especially at low flowrates.

   In any test, it should always be ensured that there are no pockets of trapped air in the test section which are subject to variable pressures, before, during or after the test.

4. Once the air from the pipe has been eliminated, the regulation valve can be adjusted until the desired flowrate is reached. It is important to remind at this point the tolerances for flowrates specified in the standards. For instance, the test at minimum flowrate (ISO 4064:1993) should be carried out at a flowrate between the minimum flowrate and 1.1 times that value. Additionally, when adjusting the flowrate, the capacity of the bench to maintain a stable flowrate should be taken into account during the time of the test. At lower flowrates, between the minimum and the transitional flowrates, the admissible variation of flowrates during the test is lower than at higher flowrates.

The adjustment of flowrates beforehand if the bench is equipped with automatic regulation valves may not be necessary. However, in this case, the regulation capacity of these valves and the response times should be carefully studied, especially in tests with short durations.

5. Once the flowrates have been set to their correct values, and the main valve situated downstream from the testing section has been closed, an initial reading can be obtained from the meters. In order to do it, the meters must have come to a complete stop and no leak should exist. The reading can be done manually or automatically, in which case the total volume would be accumulated by "steps" of the hands or other indicating devices of the totalizer.

   This step is extremely important, since the metering error is calculated from the reading of the meters at the beginning and the end of the test. Consequently, special care should be taken to avoid introducing errors at this stage.

   The most common faults when reading a meter are:

- The totalizer rollers (with the numbers) and the hand are not completely synchronized. For instance, the hand may be signalling zero, while the roller showing the litres is only half on its way (Figure 8.6).
- Lack of alignment between the hands. This error may occur in the cases where the meter only displays the volume through hands.
- Mistakes in the registration of the reading. Although the visual reading of the meter may be correct, when it is noted down on paper or on the computer, the numbers are changed in order or introduced incorrectly.

FIGURE 8.6. LACK OF SYNCHRONIZATION BETWEEN THE WHEELS AND THE HANDS OF THE TOTALIZER

- The frequency response of the optical reader or the data acquisition equipment is not high enough to register changes at the correct pace. Data are lost, generating an under-metering error which does not correspond with reality.

6. One final check is necessary before beginning the test. The control device should be reset and be ready in order to begin registering flow during the test.

7. Once everything is ready the main valve is opened. The operation of this valve (both at opening and closure) should be fast enough to avoid introducing uncertainties in the tests, but slow enough to avoid generating high or low pressures. Full details on valve operation times can be found in several standards, such as OIML R49, ISO 4064 and EN 14154-3. Should there be doubts about whether if the operation time of the valve affects the results of the tests, it is recommended that the tests should be made longer, and never under 60 seconds.

    The stopwatch should be started at the time of opening the valve in order to get a value for the average real flowrate during the test.

8. During the test, the circulating flowrate should be checked to ensure that its value corresponds to the desired value and it is not deviated more than a 2.5% (when the test flowrate is between the minimum and the transitional values) or 5% (for values between the transitional and maximum flowrates). This is especially important at low flowrates, where the variation of the error curve with the flowrate is very steep and any small variation can completely modify the meter's error.

9. Once the target volume for the test has been reached, the main valve is closed. This operation can be done manually, although it is recommendable to perform it automatically triggered by a sensor. At this point the stopwatch should also be stopped.

10. The meter is read once it comes to a complete stop with the highest possible resolution. Special care should once again be taken to ensure a proper reading.

11. The meter's error is calculated (Equation (8.2)) from the accumulated volume by the control device and the two readings of the meter, at the beginning and end of the test. The flowrate that will be associated to that error will be obtained (Equation (8.3)) from the division of the total accumulated volume in the reference device and the time registered by the stopwatch:

$$\varepsilon = \frac{(\text{initial reading} - \text{final reading}) - \text{control device accumulated volume}}{\text{control device accumulated volume}}$$

$$(8.2)$$

$$\text{Testing flow rate} = \frac{\text{control device accumulated volume}}{\text{test duration}} \qquad (8.3)$$

8.4.2. Testing procedure of a meter with the flying start and finish method (EN 14154-3)

The basic precautions that must be taken when testing the meter following the flying start and finish procedure are similar to those described in the previous test methodology. However, in the case of the flying start–finish, the testing section has an additional element, the flow diverter, which basically acts as a three-way valve, diverting the flow to a drain or to the reference tank:

1. As in the previous case, the meters should be installed so that the flow is in the correct direction. The distance between the different meters in the test bench should also be enough to ensure that the upstream meter does not interfere with the meter placed downstream.

   Before any water is circulated through the testing section the flow diverter should be positioned towards the drain.

2. The purge of the testing section should be performed without exceeding the maximum flowrate or the "normal" operating flowrate for meters retired from the field.

   Once the air has been eliminated from the pipe and meters, the test flowrate can be adjusted by means of the regulating valve or by changing the speed of the pumps. In order to guarantee the reliability of the tests, the flowrate should be stabilized for a reasonable time before the start.

3. The reading of the accumulated volume of the meter in this procedure must be done electronically. For this purpose, the corresponding sensors should be placed on the totalizers in order to capture the flow signals provided by this element.

   A distinction should be made between continuous signals, which provide information at all times on the circulating flowrate; for instance, an analogue signal of 4–20 mA from an electromagnetic meter, and digital signals in which a pulse is generated every time a certain amount of volume circulates through the meter, created by a pulse emitter or the low flow indicator of the meter.

   In any case, and depending on the type of signal extracted from the meter, a valid integration procedure should be implemented during the duration of the test.

4. The signal from the meter should be coordinated with the operation of the flow diverter. For instance, if a pulse emitter is used to read the meter, the flow diverter must not divert the flow to the tank until the signal from the pulse emitter has changed its state. Otherwise, the volume circulated and diverted during the first pulse would be different from the theoretical volume of the pulse and errors on the calibration procedure will be generated.

5. The manoeuvre of the flow diverter must be symmetrical, fast enough, repetitive and independent of the circulating flowrate. One of the main sources of uncertainty in the test is originated in the movement of the flow diverter. Usually, the stopwatch is started when the flow diverter passes through the intermediate position between its two final states.

6. During the test, the flowrate should be kept stable, with variations within the specified limits.

7. The test can be ended in two ways. If the signal from the meter is analogue, or a high-frequency digital signal, once the circulated volume reaches a certain value the test can be considered ended, signalling the flow diverter to change its position.

   If the signal is a low-frequency digital one, for instance a pulse emitter, the test must continue until the signal changes its state. Only then can the flow diverter be signalled to change its position. In such cases, and when using a calibrated tank, the tank should be prepared to provide readings on a wide range of volumes, in order to avoid that in the meantime the circulated volume exceeds the admissible range for the tank.

8. Once again the flow diverter should operate quickly and independently of the circulating flowrate, and the stopwatch will be stopped when the diverter passes through the intermediate position between its two final states.

9. The metering error and the test flowrate will be obtained exactly like in the previous case.

Although this methodology is more complex due to the necessary electronic instrumentation, the reliability of the results is higher than the one obtained with stopped meter tests. There are basically two reasons. Firstly, the flowrate is constant before, during and after the test, eliminating the influence of inertias and working with the meter always at the same operating point. Additionally, the reading of the meter is performed electronically, avoiding the quite usual human error.

8.5. TESTING FOR ERROR CURVE VERIFICATION ON THE FIELD

Whenever a complaint is received from a user, it is convenient to test the meter at the user's facilities. This allows:

- Saving time, since the test is performed on a single visit and there is no need to return the meter to the user after the test.
- Reducing costs. It is not necessary to remove the meter to test it in a laboratory and to install a replacement meter. Additionally, if the replacement meter is not a final one, additional costs will derive from installing back the original meter.
- Identifying possible causes of metering error in the installation that would not appear in a laboratory test.

However, the uncertainty associated with these tests is much higher than the one attainable at a laboratory. It should be taken into account that the different factors influencing the test are much more difficult to control under these circumstances. Additionally, the reference devices, usually flow meters or special meters, have a higher associated uncertainty than the devices used in the laboratory. Finally, the testing procedure may be complicated under certain specific circumstances, adding further uncertainty to the whole test.

8.5.1. Equipment used for field testing of meters

The only requisite in testing the error curve of a meter in the field is to use a valid reference device to compare the meter reading. Most manufacturers can provide such a device, or even the utility can have one tailor made for its necessities. Figure 8.7 shows an example of the basic schematics for such a setup.

The necessary equipment for a basic field test of a meter is a high-quality flow or water meter acting as a reference instrument, one or several regulation valves, one or several shut-off valves and a connection hose or pipe.

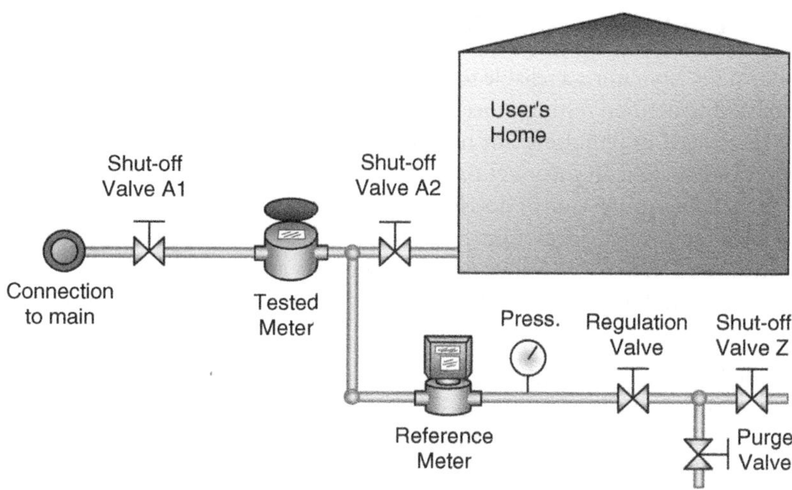

FIGURE 8.7. SETUP OF TESTING EQUIPMENT WITH A REFERENCE METER FOR FIELD USE

This setup can present multiple variations. In any case, it should be taken into account that:

- The connection to the pipe should be such that it is perfectly watertight and is easy to install and uninstall.
- The manometer should be used to guarantee that the system is watertight. The tested meter should never be used to guarantee that no water is lost. The resolution of the manometer should be able to detect pressure variations of 0.1 bar. It is not recommendable, despite being quite common, to use low-quality manometers with a very small sphere diameter.
- The reference meter should provide information on the instantaneous flowrate in order to adjust the test flowrates accordingly. Additionally, it should be able to totalize the volume circulated during the test. The reset procedure for the reference device should be simple and be performed before every test, guaranteeing that only the volume corresponding to such test is registered. The screen or display should be easily readable, with no room for mistakes and with an excellent resolution of the totalized volume. Most digital displays fulfil these requirements.

    The reference device should be often calibrated at the laboratory and the error curve known to perfection at least at 10 points of the flow range. The test conditions should be reproduced in the laboratory as faithfully as possible in order to determine the real performance of the complete device during the field tests.
- The regulation valve enables the adjustment of the flowrate during the test. In some cases, and depending on the range of flowrates of the equipment, it may be necessary to use different valves for low flow and for medium and high flowrates.
- The shut-off valve, usually a ball valve or an electronically controlled valve, must guarantee a quick operation and avoid losses.

8.5.2. Calibration procedure

Just like in the laboratory, a reliable test of a water meter on the field requires a perfectly defined and structured testing procedure. Usually, the staff performing these tests are very experienced in plumbing but not in device calibration and is not fully aware of the implications that an incorrect procedure may have on the final results.

It is therefore important to develop instructive programmes in which the testing procedures are covered from a practical point of view, pointing out the most common errors and stressing the objectives that the tests are trying to achieve.

Generally speaking a testing procedure should cover the following aspects:

1. Prior to the test it is necessary to know the behaviour of the reference device throughout the measuring range. In order to achieve this, the equipment should be calibrated as often as necessary, for instance once a month. When possible, such calibrations should be performed with the whole equivalent field setup. This is especially important to check whether low-pressure values, close to the atmospheric pressure, affect the performance of the meter. Ideally speaking, the pressure in the manometer should never drop below 0.6 bar. Additionally, all valves should be watertight and so should be all internal connections in the calibration equipment.
2. Connect the calibration equipment to the pipe where the meter is located. The further away this connection is from the tested meter, the higher the uncertainty of the test will be. Once again, all the equipment and connections should be checked for losses, even the smallest ones.
3. Use the manometer to check whether the system is truly watertight. In Figure 8.7 the verification device is placed immediately downstream from the meter, while in Figure 8.8 it is located inside the users' facilities. It is obvious that the second

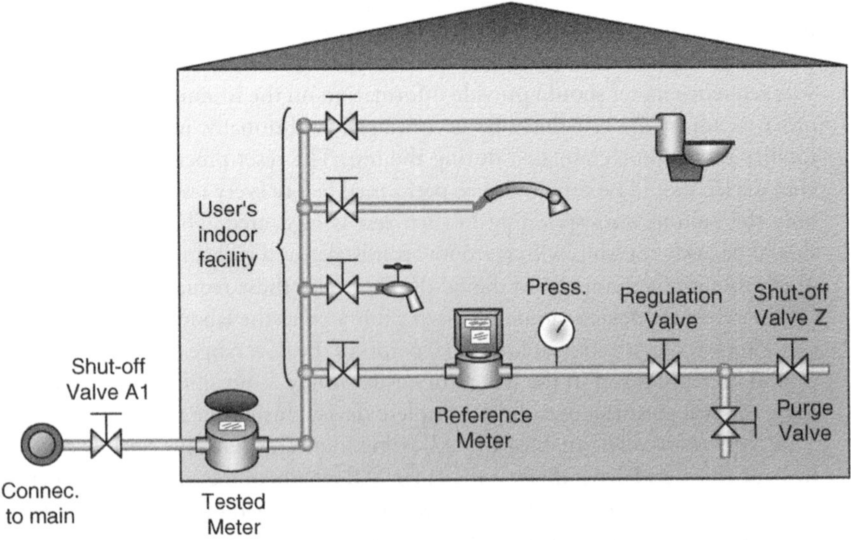

FIGURE 8.8. ALTERNATIVE LOCATION OF THE VERIFICATION EQUIPMENT

option presents many problems compared to the first one, at least regarding the uncertainty associated to the tests. After all, the final objective is to achieve that all the volume that has circulated through the tested meter goes through the reference device as well.

In Figure 8.7, if once the verification equipment has been connected and all valves have been closed (A1, A2 and Z) the pressure drops, it may be due to a leak in the connections, in valve A2 (valve Z was supposedly tested in the laboratory) or at the users' facility. If the pressure remains constant, the purge valve must be opened slightly and very slowly, allowing a minimum flowrate and verifying a pressure drop. Otherwise, it is possible that valve A1 is not completely watertight and then it would not be possible to properly check valve A2. In this case, it is not possible to guarantee that the entire flowrate during the test circulates through the reference device.

The procedure to follow in Figure 8.8 is similar except for the need to check all valves in the installation. First, all valves in the system and the verification equipment are closed and the pressure is monitored to check if it drops. A change in pressure would mean there is a leak and it needs to be located. Otherwise, the purge valve can be opened slightly and very slowly, with a target flowrate of 10 l/h. If the pressure is then maintained constant, the installation will not be suitable to test meters. Two are the main reasons for this. Firstly, it is possible that valve A1 is not completely watertight, and it is not possible to check the state of the other valves. Secondly, it is possible that the pressure is kept constant for a while due to the presence of hydraulic accumulators in the user's facility. In this case, it would not be recommended to test meters in this installation.

4. Once the users' facility has been verified, it is necessary to adjust the test flowrate. With the shut-off valve open, the regulation valve is operated. Once the desired flowrate is reached, and there is confirmation that it will be maintained constant throughout the test, it is admissible to close the shut-off valve Z.

5. The totalizer is then reset, and the tested meter is read. It is important to take into account the minimum resolution of the meter and the test volumes. The test volumes should be at least 100 times larger than the minimum resolution of the meter. Otherwise, the uncertainty associated with the test will be too large, greater than 2%.

6. Shut-off valve Z is opened and the stopwatch is started.

7. When the reference meter has registered approximately the volume allocated for the test, or if the tested meter shows no signs of registering, valve Z is closed and the stopwatch is stopped.

8. Both the tested and the reference meters are read.

9. The metering error of the tested meter at the specified flowrate is calculated by:

$$\varepsilon = \frac{(\forall_i)(1 + \varepsilon_{\text{reference}})}{\forall_{\text{reference}}} - 1 \qquad (8.4)$$

where:

$\forall_i$ is the volume indicated by the meter to be calibrated, that is the difference between the initial and final readings.

$\forall_{\text{reference}}$ is the volume indicated by the reference meter.

$\varepsilon_{\text{reference}}$ is the relative error of the reference meter at the tested flowrate (p.u.).

TABLE 8.4. STANDARD UNCERTAINTY OF A METER AS A FUNCTION OF THE SCALE RESOLUTION AND THE TOTAL VOLUME CIRCULATED IN THE TEST (TWO READINGS)

		Test volume (l)							
		1	5	10	20	50	100	200	500
Scale	0.025	1.021%	0.204%	0.102%	0.051%	0.020%	0.010%	0.005%	0.002%
resolution	0.05	2.041%	0.408%	0.204%	0.102%	0.041%	0.020%	0.010%	0.004%
of the	0.1	4.082%	0.816%	0.408%	0.204%	0.082%	0.041%	0.020%	0.008%
meter (l)	0.15	6.124%	1.225%	0.612%	0.306%	0.122%	0.061%	0.031%	0.012%
	0.2	8.165%	1.633%	0.816%	0.408%	0.163%	0.082%	0.041%	0.016%
	0.5	20.41%	4.082%	2.041%	1.021%	0.408%	0.204%	0.102%	0.041%
	1	40.82%	8.165%	4.082%	2.041%	0.816%	0.408%	0.204%	0.082%

The actual test flowrate can be obtained from several readings of the reference device during the test (if it provides such information) or by:

$$Q = \frac{3600 \cdot \forall_{reference}(1)}{(1 + \varepsilon_{reference}) \cdot t_{i-f}(s)} \tag{8.5}$$

where $t_{i\text{-}f}$ is the duration of the test.

Regarding the testing procedure there are a few additional details that require further attention.

Firstly, and depending on the characteristics of the devices there must be a target uncertainty for the test. The uncertainty associated with the reference device should be in the range of 0.5% of the measure. A reasonable target value for the overall uncertainty of the test would be 1% or 1.5% of the measure.

As explained previously, the total volume circulating during the test affects the global uncertainty of the test (Table 8.4). It can be shown that if the uncertainty associated with the meter reading is expected to be maintained in values lower than 0.5% when the reading resolution is 0.2 l, which is the maximum admissible value according to ISO 4064:1993 for a meter with a nominal flowrate of $1.5 \, m^3/h$, then the circulated volume needs to be at least 20 l. If the test is carried out at 30 l/h, the shortest required test time is 40 minutes.

Consequently, the volume to be circulated will depend on the target uncertainty and the resolution of the meters, for both the test and the reference meter.

Additionally, it is also important to take into account that any kind of leak in the valves will bias the results. For that reason it is critical to guarantee that the shut-off valves at the users' facilities and testing bench are completely watertight. Under no circumstances this warranty can come from the reading of the tested meter with unknown low flow sensitivity.

The following example shows the impact of a leak in the estimated metering error when a leak is present and valves are not completely watertight (Figure 8.9). Imagine the ideal case where both, the tested and the reference meter, have no error of indication at a flowrate of 300 l/h. If then the meter is tested with a small leak of 10 l/h, this

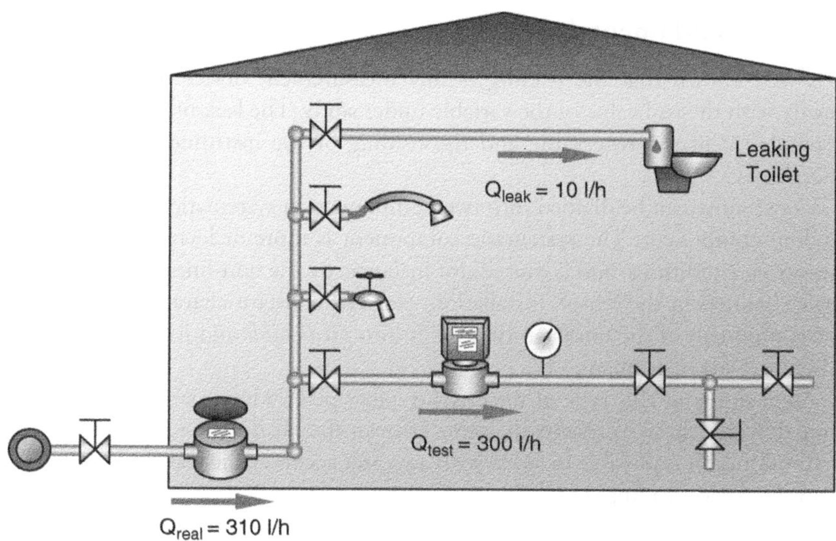

FIGURE 8.9. IMPACT OF A LEAK IN THE CIRCULATING FLOWRATES IN A TEST

would produce an error for the tested meter of $+3.3\%$, as opposed to the 0% which the test would have produced without a leak. If the meter is tested at 30 l/h instead of 300 l/h then the calculated error would go up to $+33.3\%$.

As a conclusion it is important to point out that, unless the valves are completely free from leaks, the results will be biased towards positive errors, and their magnitude will be bigger with smaller test flowrates.

8.6. UNCERTAINTY ANALYSIS DURING TESTS

Many of the tests performed both in the laboratory and on the field are carried out after a complaint has been filed by users. As a consequence, the assessment of the meters' error has to be reliable enough to advice which further actions need to be taken according to them.

Although field tests can never be as reliable as laboratory tests, they still need to deliver a low uncertainty. Consider the case where a utility needs to inform a user that the error of his/her meter is -3% but there is a certain probability for it to range from $+1\%$ to -7%. This test would not be useful despite the considerable economic cost, and it would additionally generate further doubts and mistrust in the user. Consequently, and depending on the circumstances, sometimes no test is better than a defective one.

In any case, regardless of whether the tests are for the internal use of the utility or to satisfy a user request, it is always of great importance to control the uncertainty associated to the error tests. This uncertainty will be a function of the procedure and the materials used for the test. This section provides guidelines to assess this parameter.

8.6.1. Fundamentals

It is well known that the reading of any measurement device never corresponds exactly with the real value of the variable under study. The lack of *security* in the correspondence of the true value and the reading of the instrument constitutes the uncertainty.

Uncertainty can be divided into two components: a systematic component and a random component. The systematic component is more or less constant for given measuring conditions, and is caused, for instance, by the non-linear behaviour of the device, failures in the sensor, installation, etc. The constant character of the systematic component of the uncertainty turns it into *predictable* and it can be reduced by calibrating the instrument.

An example of this type of uncertainty takes place when a certain meter, at the same flowrate, always presents the same error in the measurement. By knowing this circumstance, it is possible to assess with a greater accuracy the real circulated volume from the device reading.

On the other hand, random uncertainty has variable magnitude and sign, and cannot be eliminated by calibration. However, it is true that random errors with constant metering conditions oscillate around zero. Consequently, with a large number of measurements and averaging them, the impact of this uncertainty can be reduced. In the ideal case of an infinite number of readings taken, the average of random errors would be zero.

However, in practice it is not possible to take a large number of readings, mainly because often it is not possible to guarantee that the system conditions will remain stable with time. In this case, part of the random character of the uncertainty is also produced by the variability of the test conditions.

8.6.2. Calculating uncertainty in the determination of a meter's error

Assessing the uncertainty while determining a meter's error at a certain flowrate may turn up to be an extremely complex process due to the high number of factors that may influence the test results. The following are simplified guidelines on how to assess this parameter and the possible sources of uncertainty:

1. Uncertainty due to the reference device.
   The tested meter's error is obtained by comparison of its measurements with those of a reference device, either a tank of known volume, a scale or a reference meter. In any of those cases, the readings of such devices are subject to error. The uncertainty of the reference device is the discrepancy range between the measurement and reality with a certain associated probability. Typical values are 0.05% for a tank or a scale, or 0.5% for a reference meter. Following up, for calculation purposes, this value will be called u_0. The uncertainty of the reference device can usually be obtained from its calibration certificate.

2. Uncertainty due to the reading resolution of the tested meter.
   The volume measured by the tested meter is obtained from the difference in readings before and after the test. Consequently, the value of the total volume will depend

on its reading resolution. In the worst possible scenario, the error in each reading will be half of that resolution. In such case, the uncertainty associated with a single reading would be calculated by:

$$u_{\mathrm{c}} = \frac{d_{\mathrm{c}}}{2\sqrt{3}\forall_{\mathrm{test}}} \quad \text{(one reading)} \tag{8.6}$$

where d_{c} is the resolution of the tested meter and $\forall_{\mathrm{Test}}$ is the total circulated volume.

However and as mentioned above, in order to obtain the error, the value of the total volume measured is needed. In most cases, such volume is also calculated from the difference of two readings, and consequently the uncertainty is determined by:

$$u_{\mathrm{c}} = \frac{d_{\mathrm{c}}}{\sqrt{2}\sqrt{3}\forall_{\mathrm{test}}} \quad \text{(two readings)} \tag{8.7}$$

Table 8.4 shows the standard uncertainty associated to a meter as a function of its resolution and the total volume of water circulated during the test. In this table, it has been assumed that the total volume has been calculated from two readings. If such value had been obtained from a single reading (for instance from a meter with a reset device) the percentages in the table should be multiplied by 0.707.

3. When a reference meter is used for the test, the uncertainty associated to the reading of the device (u_{p}) will be obtained in the same way. If the reference device is not a meter, the resolution of the device should be used instead of the meter's resolution.

4. If the test flowrate cannot be kept constant, an uncertainty factor appears due to lack of stability of the flowrate. This component may become important at low flowrates, especially below the transitional flowrate. The international standards require for the test to be valid, between the minimum and transitional flowrates, that the flowrate, once its stationary value has been reached, remains within 2.5% of such value. If the test is carried out at higher flowrates, a 5% variation is admitted.

5. Another important source of uncertainty is associated to the real flowrate. Quite often, in order to simplify data analysis, the tests are associated to the nominal flowrate that was defined in the testing protocol. However, when the test takes place, although the real flowrate is adjusted to match this value as closely as possible, the real value may differ. For instance, the protocol may state that the test must be carried out at 30 l/h, however in practice, it is impossible to set the real flowrate, which is obtained by operation of the regulation valve, at that precise value.

The error curve changes quite steeply at low flowrates, and a deviation of 2 or 3 l/h may create significant differences in the results. In practice, the average real value of the flowrate during the test is not known, and even if it was obtained, it would complicate the analysis of the results excessively. As a consequence, certain uncertainty is generated when assessing the overall error at the target flowrate.

Quite obviously, this uncertainty is different for each meter, and depends on the test flowrate, and it is therefore quite difficult to estimate beforehand. As a general rule, for flowrates higher than the transitional flowrate, this component is negligible. For flowrates ranging from the minimum to the transitional values, it will depend on the metrological degradation of the instrument. If the errors are close to zero ($\pm 5\%$) they can also be neglected. Otherwise they will have to be taken into account. Although it is convenient to determine the real order of magnitude, an approximation can be used:

$$u_{Q\_test} = 0.1 \cdot \text{Abs}(\varepsilon_{Q_{test}} \ (\text{p.u.})) \tag{8.8}$$

For example, if the error of indication at the test flowrate is -15%, the uncertainty component due to the determination of the test flowrate would be approximately 1.5%.

Finally, the estimation of the combined uncertainty, which take into account all the sources mentioned above, would be assessed by:

$$u = \sqrt{(u_0)^2 + (u_c)^2 + (u_p)^2 + (u_{Q\_test})^2} \tag{8.9}$$

Since the combined uncertainty originates in at least three contributions of magnitudes which are comparable and independent, the central limit theorem can be applied, assuming that the combined uncertainty follows a normal distribution. The expanded uncertainty would be obtained (Equation (8.10)) by multiplying the combined uncertainty by the coverage factor k, which for a desired value of the coverage probability of 95% would take a value of 2. For other probabilities refer to Table 6.4:

$$U = k \cdot u \tag{8.10}$$

In order to understand the previous calculation, suppose that a meter has been tested for a certain flowrate. The error obtained for the test is ε. Since neither the procedure nor the instruments used as a reference are perfect, it is not possible to assure that ε is truly the real error. The correct form to express this error would be "the error of the meter is ε, although with an 95% of probability (depending on the value of the coverage factor) it will take values ranging from $\varepsilon - U$ to $\varepsilon + U$.

In any case, further explanation on how to obtain the calibration uncertainty can be obtained in ISO 5168 "Measurement of fluid flow – Evaluation of uncertainties".

Example

A used meter is retired to be tested at 30, 120 and 1500 l/h. The meter has a scale resolution of 0.2 l. For the test, an electronic meter is used with an associated uncertainty of 0.5% and with a scale resolution of 0.2 l. The electronic meter has a reset bottom to bring reading to zero at the beginning of the test. The volume of water used for the tests is 10 l at 30 and 120 l/h, and 50 l at 1500 l/h. The obtained errors at 30, 120 and 1500 l/h are -12.3%, -1.6% and -0.4%, respectively.

The uncertainties associated to each of the parameters for the different test flowrates are shown in Table 8.5.

In this case, the contribution of the uncertainty associated to the reference device, u_0, is for all cases 0.5%. The uncertainty in the tested meter reading, u_c, is obtained directly from Table 8.4 as a function of the circulated volume, 10 or 50 l depending on the case, and the resolution of the meter, 0.2 l. The uncertainty due to the reading

TABLE 8.5. CALIBRATION UNCERTAINTIES OF A METER

Test (l/h)	u_0 (%)	u_c (%)	u_p (%)	u_{Q_test} (%)	u (%)	$U = k \cdot u$ (%)
30	0.50	0.816	0.577	1.23	1.66	3.32
120	0.50	0.816	0.577	0.16	1.13	2.26
1500	0.50	0.163	0.115	0.00	0.54	1.08

of the reference device, since only one reading is performed, is obtained by multiplying the corresponding values in Table 8.4, depending on the circulated volume and the resolution, by 0.707. This uncertainty is reduced by the fact that the device is equipped with a reset device that eliminates the uncertainty related to the initial reading. The uncertainty due to the real circulated volume is calculated with the following considerations:

- At 1500 l/h is neglected.
- At 120 l/h and 30 l/h is estimated as a 10% of the meter's error.

The combined uncertainty is finally established by Equation (8.9). Finally, in order to calculate the expanded uncertainty, a value of 95% has been chosen for the coverage factor k, and therefore a value of 2 is used.

The interpretation of the results would be as follows: The error obtained at 30 l/h is -12.3%. However, with a probability of 95%, the real error could range from -15.62% to -8.98%.

Quite clearly, the magnitude of the uncertainty may be a source of conflict with users and also may lead to wrong conclusions.

8.7.   REFERENCE STANDARDS

Relevant standards and technical specifications:

TC 30/SC 9

ISO 4006:1991 Measurement of fluid flow in closed conduits – Vocabulary and symbols

ISO 4185:1980 Measurement of liquid flow in closed conduits – Weighing method

ISO 4185:1980/Cor 1:1993

ISO/TR 5168:1998 Measurement of fluid flow – Evaluation of uncertainties

ISO/TR 7066-1:1997 Assessment of uncertainty in calibration and use of flow measurement devices – Part 1: Linear calibration relationships

ISO 7066-2:1988 Assessment of uncertainty in the calibration and use of flow measurement devices – Part 2: Non-linear calibration relationships

ISO 8316:1987 Measurement of liquid flow in closed conduits – Method by collection of the liquid in a volumetric tank

ISO 9368-1:1990 Measurement of liquid flow in closed conduits by the weighing method – Procedures for checking installations – Part 1: Static weighing systems

ISO 11631:1998 Measurement of fluid flow – Methods of specifying flow meter performance.

9

Quality control

9.1. INTRODUCTION

Every meter, before it leaves the factory, is checked and analysed individually. During this process, known as primitive verification, the metering error is obtained and verified at three different flowrates (minimum, transitional and maximum). However, in practice, the manipulation of the meters at a later stage (for instance during transportation) may alter the performance of the meters when finally installed at the users' facilities.

The cost of the installation of a new but defective meter depends on the price of water, the administrative and personnel costs implied in its substitution and the cost of the device itself.

Obviously, not all manufacturers supply meters of equal quality. Two-meter models of the same metrological class may be assembled in manufacturing processes where the quality controls and the production standards are very different.

The utility may choose to trust the manufacturer and its quality control procedures and accept the meters without further proof of their performance. However, the experience shows that meter shipments often contain defective meters and the percentage of faulty units changes significantly from one supplier to the other.

It is therefore convenient to test the quality of a batch of meters before they are installed. The tests to be carried out will be, in principle, the same tests carried out for the primitive verification of the meter.

Testing meters at their reception may be a costly process. If the utility chooses to check every device that is delivered the process becomes clearly inefficient. In this case, the potential benefits will probably be less than the costs implied in the operation.

For this reason, and similarly to the procedures followed in other industries, it is convenient to implicate the manufacturers in the quality assurance process. This action significantly reduces the number of tests needed when receiving a lot. In this case, the decision about the acceptance of a certain batch of meters would depend on the number of successful inspections, a figure agreed beforehand with the manufacturer. With this option, it is possible to reduce the associated costs while maintaining a good-quality control on the purchased devices.

9.2. SAMPLING PLANS

The first obvious consequence of testing a limited number of meters is that there is a certain probability of accepting a defective batch of meters and a certain chance to reject a shipment of meters that are up to the required quality standards. The probability of acceptance of a defective shipment is called a type I or α error, and may be considered a risk on the buyer's side. The probability of rejection of a correct shipment is known as a type II or β error, and defines the seller's risk.

Defining a sampling plan implies to select the desired probability of type I and II errors. The manufacturer will obviously want to minimize type II errors, while the utility will try to reduce type I errors without increasing the testing costs excessively. After all, the larger the sample the smaller the values of type I and II errors, with higher testing costs. If the sample represents the whole shipment, the error probability is 0.

There are several possibilities to define the sampling plans which are widely covered by the literature. The control and acceptance criteria may be defined observing the measurable characteristics of the acquired good, controlling the reception by variables, or classifying the elements of the batch in two categories (defective or non-defective) in which case it would be a control by attributes. In practice, the second type of control is the most extended one for economical and practical reasons.

However, regarding water meters, it is convenient to carry out a control by variables, for the error curve is after all a quantitative parameter. A control by attributes could also be performed classifying each meter in defective or valid. In such case, testing the meters is also required and any economic advantage of following this type of quality control procedure is lost.

A quality control by variables would consist in testing the meters' error curve at previously specified flowrates, for a certain sample of the shipment. Depending on the results for each flowrate, the batch would be rejected or not. The flowrates and acceptance limits could be set according to local or national regulations or by agreed with the supplier. However, as it will be seen later, in a quality control procedure by variables a batch may be rejected even in the case when the average error of the meters tested is within tolerable limits.

9.3. DEFINITIONS

Before describing the quality control process it is necessary to define some parameters and statistical terms which are commonly used.

Batch: Quantity of product delivered for inspection.

Sample: Random selection from the meter batch. A sample is characterized by the *sample's mean $\bar{x}$* and the *sample's standard deviation s*.

Acceptable quality level (AQL):   Defined in ISO 3951 as the maximum percentage of defective units that, from the sampling point of view, can be considered as satisfactory. This parameter is one of the most important concepts in statistical sampling techniques and it establishes the threshold for decision making based on the sampled batch.

Quality limit: Is the quality level that could be accepted with a certain specified probability (10% according to ISO 3951). In other words, it provides information on the minimum quality of a batch with an acceptance probability of 10% (when applying ISO 3951).

Specification limit: Defined as the value established as acceptable in the sampling, in this case the meter's error at different flowrates. It can be defined as a lower limit (L_l) when it defines a minimum value or an upper limit (L_u) for a maximum acceptable value. In the case of water meters the upper limit could be $+5\%$ and the lower limit -5% for flowrates between the minimum and the transitional values. For higher flowrates, the upper and lower limits could be $+2\%$ and -2%, respectively.

The sampling of meters requires a double limit. However, a different AQL can be established for each one of them, and it could be considered a *double combined specification limit*. This supposition considerably simplifies the results' analysis.

Constant of acceptability (k):   Its value depends on the value specified for the AQL and the sample size. This constant is used as a control statistic, defining the threshold of acceptability of the batch.

Quality control statistic (Q):   Parameter resulting from the specification limit, the sample mean and the estimation of the standard deviation of the batch. The decision on the batch is taken comparing Q with the constant of acceptability k:

Quality control statistic for the upper limit: $\quad Q_u = \dfrac{L_u - \bar{x}}{s}$ $\hfill$ (9.1)

Quality control statistic for the lower limit: $\quad Q_l = \dfrac{\bar{x} - L_l}{s}$ $\hfill$ (9.2)

Process capability index (Cp):   Defined as the ratio between the amplitude of the specification range $(L_u - L_l)$ and six times the standard deviation of the sample:

$$\mathrm{Cp} = \frac{L_u - L_l}{6 \cdot s} \tag{9.3}$$

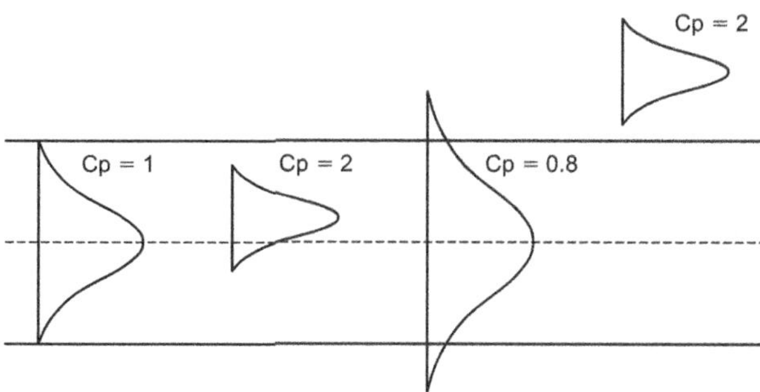

FIGURE 9.1.   **PROCESS CAPABILITY INDEX**

Values of Cp above 1 mean that at least 99.7% of the meters have an error within a range of amplitude of 10% between the minimum and the transitional flowrate and a 4% at higher flowrates. However, this parameter does not indicate whether such variability falls within the specification range, or if it is partially or totally excluded (Figure 9.1).

In general terms, the larger the value of Cp the better the manufacturing process is. A high Cp value implies that the manufacturer has a strict control of the production process and all the meters come out consistently with the same performance values (regardless of whether such performance is a good or a bad one).

Sometimes, a low value of Cp is not necessarily caused by the poor quality of the meters but instead by an inadequate testing procedure. Quality engineers need to be aware of the possible origins for the variability of the sample.

Process performance index (Cpk):   This parameter reflects the situation of the error distribution within the preset specification limits. It is defined by:

$$\text{Cpk} = \text{minimum} \left| \left(\frac{\bar{x} - Li}{3s} \right) \leftrightarrow \left(\frac{L_s - \bar{x}}{3s} \right) \right| \tag{9.4}$$

A value of Cpk above 1 means that a high percentage (>99%) of the production falls within the specification limits. If the value of Cpk is below 1 it would mean that a significant percentage of the batch is outside the limits. A value of 0 appears when the sample mean is coincident with one of the specification limits, and as a consequence 50% of the meters would present errors below or above such limits. When the sample mean is outside the specification limits, the Cpk presents negative values (Figure 9.2).

9.4.   DESCRIPTION OF THE QUALITY CONTROL PROCEDURE

The first step to define a sampling procedure is to define the AQL. This parameter specifies the number of defective units which are acceptable per batch. Although any value can be selected, it is recommendable to select the values specified by the ISO 3951

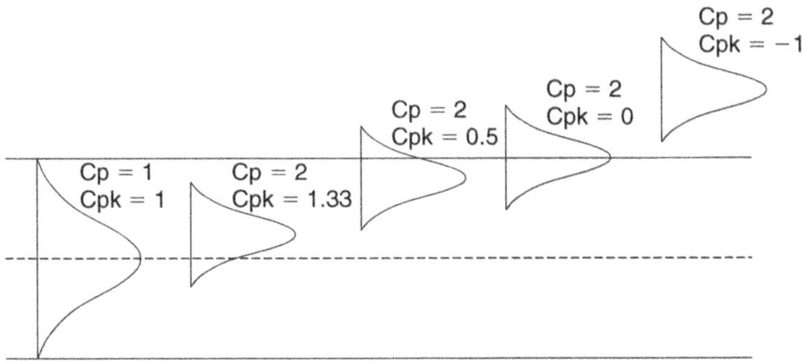

FIGURE 9.2. PROCESS PERFORMANCE INDEX

TABLE 9.1. LETTER CODE FOR BATCH SIZES AND INSPECTION LEVELS

Batch size	Inspection level		
	I	II	III
151–280	F	G	I
281–400	G	H	J
401–500	G	I	J
501–1200	H	J	K
1201–3200	I	K	L
3201–10,000	J	L	M

standard and to conform to it. Since the standard deviation for the metering error at different flowrates, σ, is not known, it will be necessary to apply the "s" or the "R" methods. Both are described in detail in the aforementioned standard.

Three levels of inspection are possible: reduced (level I), normal (level II) or rigorous (level III). A rigorous inspection plan will be applied for non-reliable suppliers. If a supplier has proven to have a quality production and systematically pass all controls, a reduced inspection plan will be applicable. In all other circumstances a normal plan will be applied.

According to ISO 3951, a letter code is to be assigned to every batch size and level of inspection to identify the inspection plan. The codes are shown in Table 9.1.

For the "s" method, each code letter has a corresponding sample size, independent of the AQL (Table 9.2).

Once the code letter of the sampling plan is known, together with its level of inspection, Tables 9.3–9.5 provide the values of the constant of acceptability k, which plays the role of quality control statistic to accept or reject a shipment.

Additionally, it is necessary to define the specification limits. For the quality control of new water meters these will be of ±5% between the minimum and the transitional flowrate and ±2% for flowrates equal or above the transitional value. With such limits it is possible to obtain the quality control statistic Q. This parameter, for the upper and lower specification limits is calculated with Equations (9.1) and (9.2).

TABLE 9.2. BATCH SIZE FOR THE "S" METHOD

Code	Sample size
F	10
G	15
H	20
I	25
J	35
K	50
L	75
M	100

TABLE 9.3. k VALUES FOR A REDUCED INSPECTION LEVEL: "S" METHOD

Code	AQL: reduced inspection (level I)					
	0.25%	0.40%	0.65%	1.00%	1.50%	2.50%
F	1.88	1.65	1.45	1.34	1.17	1.01
G	1.88	1.65	1.53	1.40	1.24	1.07
H	1.88	1.75	1.62	1.50	1.33	1.15
I	1.98	1.84	1.72	1.58	1.41	1.23
J	2.06	1.91	1.79	1.65	1.47	1.30
K	2.11	1.96	1.82	1.69	1.51	1.33
L	2.14	1.98	1.85	1.72	1.53	1.35
M	2.18	2.03	1.89	1.76	1.57	1.39

TABLE 9.4. k VALUES FOR A NORMAL INSPECTION LEVEL: "S" METHOD

Code	AQL: normal inspection (level II)					
	0.25%	0.40%	0.65%	1.00%	1.50%	2.50%
F	2.11	1.98	1.84	1.72	1.58	1.41
G	2.20	2.06	1.91	1.79	1.65	1.47
H	2.24	2.11	1.96	1.82	1.69	1.51
I	2.26	2.14	1.98	1.85	1.72	1.53
J	2.31	2.18	2.03	1.89	1.76	1.57
K	2.35	2.22	2.08	1.93	1.80	1.61
L	2.41	2.27	2.12	1.98	1.84	1.65
M	2.43	2.29	2.14	2.00	1.86	1.67

Comparing the quality control statistics with the acceptability constant (Tables 9.3–9.5) it is possible to make a decision on the batch. Specifically, if Q_u and Q_l are greater or equal than k, the batch would be accepted. However, if either Q_u or Q_l are below the value of k, the batch would be rejected.

TABLE 9.5. *k* VALUES FOR A RIGOROUS INSPECTION LEVEL: "S" METHOD

Code	AQL: rigorous inspection (level III)					
	0.25%	0.40%	0.65%	1.00%	1.50%	2.50%
F	2.24	2.11	1.98	1.84	1.72	1.58
G	2.32	2.20	2.06	1.91	1.79	1.65
H	2.36	2.24	2.11	1.96	1.82	1.69
I	2.40	2.26	2.14	1.98	1.85	1.72
J	2.45	2.31	2.18	2.03	1.89	1.76
K	2.50	2.35	2.22	2.08	1.93	1.80
L	2.55	2.41	2.27	2.12	1.98	1.84
M	2.58	2.43	2.29	2.14	2.00	1.86

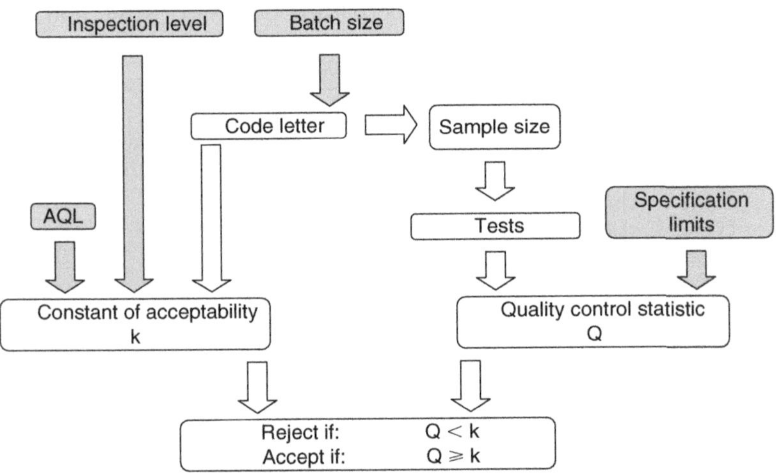

FIGURE 9.3. METHODOLOGY FOR A SAMPLING PLAN BY ATTRIBUTES ACCORDING TO ISO 3951 ("S" METHOD)

Additionally, the AQL may be different for each specification limit (upper and lower). In such case, the comparison of each quality control statistic would be referred to the corresponding constant of acceptability value.

Figure 9.3 shows a flow diagram of the methodology for the quality control by attributes of water meters. The parameters in grey correspond to initial data and their values are fixed by the characteristics of the batch (size) and the requirements of the buyer (AQL, level of inspection, limits of specification).

9.5. APPLICATION TO THE RECEPTION OF METER BATCHES

In order to better illustrate the correct procedure to perform the reception of meter batches, a practical example is described in the following pages.

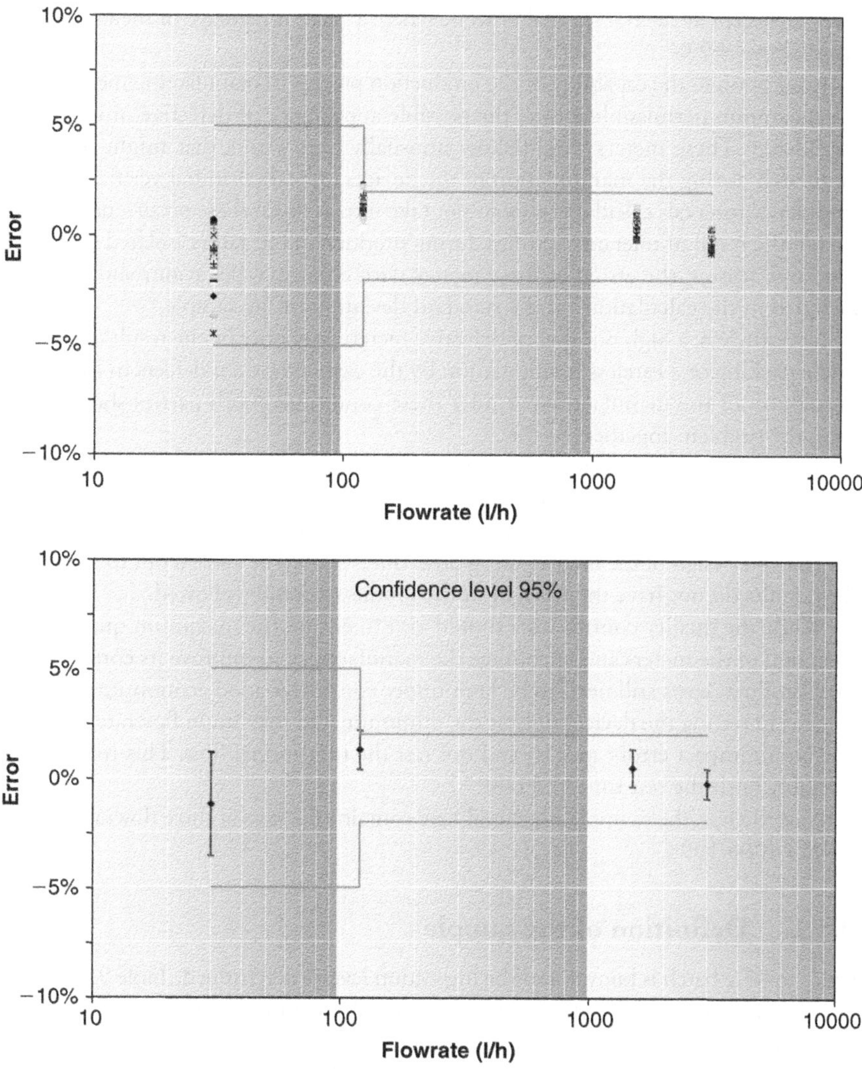

FIGURE 9.4. AVERAGE ERROR CURVE FOR 35 15 MM CLASS B METERS: MODEL 1

Water meters are mechanical instruments whose accuracy relies, in most cases, on dimensional tolerances and the quality in the finish of the different elements composing them. Despite the regulation mechanisms included, the devices never leave the factory with the same error curve. The error at each flowrate is a random variable, with a mean and a standard deviation, which is different for each meter model and manufacturing process.

Consequently, and for each batch received, an average error for every tested flowrate is obtained. This will allow estimating with certain reliability the range of values that the meter errors will have at that specific flowrate. For instance, Figure 9.4 shows that the metering error of 95% of the meters of the batch will be within the allowed error interval, except at the transitional flowrate where a small percentage of meters will have errors above +2%.

In addition to the capability of the production process to manufacture meters within the maximum permissible errors, the possible appearance of defective units must be considered. These meters would show unusually high errors that might have been originated during the manipulation of the devices after their initial verification. The confidence intervals calculated should not take into account these occurrences, as they are not the result of tolerance settings during production but rather isolated events. For this same reason, the errors of these meters with abnormal behaviour should not be included in the calculations of the standard deviation of the sample.

Nevertheless, a high number of defective meters may only be the result of inappropriate packing or a careless manipulation by the carrier, or even defects in the control procedures of the manufacturer. Under these conditions, both parties should try to solve the problem together.

Some of the most common failures found in new meters are the increase of the starting flow, high metering errors at low flowrates, normally at the minimum flowrate, and the uncoupling of the turbine and totalizer at the maximum flowrate. However, a strict control and design of the meter testing procedures should be carried out to ensure that these faults do not have their origin in the modus operandi followed.

While the quality control tests should aim to ensure the maximum quality at the reception of the meters and encourage the manufacturer to improve its control on the final product, costs still need to be kept under control. A good economical approach consists in testing the devices only at the minimum and maximum flowrate (when the maximum impact errors appear) and not test the transitional flow. This reduces considerably both the test time and cost.

Nevertheless, the example presented here contains the tests at three flowrates defined in ISO 4064:1993.

9.5.1. Definition of the sample

Once the size batch is known and the inspection level is determined, Table 9.1 will provide the letter code. With the code and Table 9.2, the sample size for the "s" method can be established.

Supposing a shipment of 1000 m and a level of inspection II, the corresponding code letter would be J. For this code letter the associated sample size is 35 units that would be randomly selected from the batch to be tested.

9.5.2. Error tests

Once the meters to be tested are selected, they should be transported to the laboratory under the usual transportation conditions of the supplier.

Figure 9.4 shows the dispersion of errors at different flowrates. As it can be seen it is much higher at low flowrates than at high flowrates where the results for different meters are quite similar.

As mentioned before, the testing procedure should be clearly specified beforehand to guarantee an adequate repeatability of the tests and to identify possible defects in the procedures. When the tests are carried out at low flowrates, special care is needed due to the fast changes of the error curve and the need to precisely adjust the test flowrate. When the flowrates are close to the minimum flowrate, a deviation of 2 or 3 l/h in the adjustment of the test flowrate may deem the meter defective and out of the allowed range. Further details on how a meter should be tested can be found in Chapter 8, which deals with testing procedures.

9.5.3. Acceptance procedures for the batch

From the error tests at each flowrate, the mean value and the standard deviation of the sample are obtained. Considering the data of our example, Table 9.6 summarizes the results of testing 35 domestic Class B meters of two different models.

The quality control statistic for each of the analysed flowrates is given in Table 9.7.

The constant of acceptability can be obtained from Table 9.4, which would correspond to a normal inspection level. In this case, for a letter code J and supposing an AQL of 0.4 (0.4% of defects), its value is 2.18.

Comparing this value with the quality control statistics obtained in Table 9.7, the table values are all higher than 2.18 except for Q_u at 120 l/h for both models (1.47 and 1.11, respectively) and also for Q_u at 30 l/h for Model 2 (1.59). This circumstance should advice the rejection of both batches, although the average value of the meters' error remains within the specification limits.

TABLE 9.6. SUMMARY OF THE ERROR CURVE TESTS

Model 1				
Sample size	35			
Nominal flowrate (l/h)	1500			
Flowrate (l/h)	30	120	1500	3000
Mean (%)	-1.15	1.34	0.53	-0.19
Standard deviation (%)	1.18	0.45	0.42	0.34
Model 2				
Sample size	35			
Nominal flowrate (l/h)	1500			
Flowrate (l/h)	30	120	1500	3000
Mean (%)	3.2	1.1	0.4	-0.1
Standard deviation (%)	1.14	0.83	0.51	0.59

TABLE 9.7. QUALITY CONTROL STATISTICS FOR THE TESTED METERS

		30 l/h	120 l/h	1500 l/h	3000 l/h
Model 1	Q_u	5.19	1.47	3.47	6.52
	Q_l	3.25	7.50	5.95	5.39
Model 2	Q_u	1.59	1.11	3.12	3.16
	Q_l	7.17	3.68	4.69	3.60

Model 2 should be rejected because the average error at 30 l/h is too close to the upper specification limit considering the value of the standard deviation.

In any case, this circumstance would not excessively affect the performance from the utility or the users' point of view, and the batch could be accepted without economic losses for the involved parties. It should be taken into account that the metering error usually evolves towards negative values, and in a short period of time the average error would certainly fall within the specification limits. A very different matter would arise from a similar situation with the lower limit, in which case it would be necessary to reject the batch.

9.5.4. Additional quality control statistics for the batch

Although the "s" method concludes once the main decision has been taken (acceptance or rejection), some further comments can be made concerning additional quality control statistics. These statistics are particularly useful when the batch is accepted. The values of Cp and Cpk for the meters tested in the example are shown in Table 9.8. These values are useful to decide which option, between those acceptable, is the best a priori.

When the value of Cp is higher than 1, the manufacturer is capable of producing 99% of the meters with errors varying less than 10% (for flowrates between the minimum and the transitional values) and 4% (for flowrates higher than the transitional value).

Cp values for Model 1 meters are higher than 1, meaning that this model is manufactured with dispersion values lower than the requested ones. Model 2 presents a Cp value of 0.80 for the transitional flowrate. This value means that even with an average error at that flowrate of 0%, some of the meters of the batch will exceed the specification limits. The lower the Cp value the higher the probability of meters in the batch exceeding the limits.

The Cpk on the other hand shows the limitations of both models, starting at 120 l/h for Model 1 and 30 and 120 l/h for Model 2. In this case, part of the production would fall out of the specification limits (for the Cpk values are lower than 1).

9.5.5. Additional considerations

The specification limits for the tests may be changed depending on the uncertainty associated to the meter testing. As described in Chapter 8 this parameter is a function of the equipment and procedures used to test the meter and of the meter itself. The

TABLE 9.8.   PROCESS PERFORMANCE AND CAPABILITY INDEXES FOR THE TESTED METERS

	Flowrate Specification limits	30 l/h ±5	120 l/h ±2	1500 l/h ±2	3000 l/h ±2
Model 1	Cp	1.41	1.48	1.59	1.96
	Cpk	1.09	0.49	1.17	1.77
Model 2	Cp	1.46	0.80	1.31	1.13
	Cpk	0.53	0.36	1.05	1.07

reason why these limits may be altered from the ones specified in the standards is that when a meter is tested, there is a doubt (*uncertainty*) about the real value of the error of that meter. When the uncertainty associated to the tests is high, the specification limits (e.g. 5% and 2%) should be "relaxed". Otherwise it will not be possible to determine if the errors are out of the limits because of the testing procedure or because of the meter performance.

Additionally, it should be taken into account that the quality of a meter should not be measured by its initial errors. Testing new meters provides information on the manufacturer's ability to produce devices within the specified limits. However, considering that all manufacturers perform an individual metrological control to each device before it leaves the factory, differences should not be significant. Discrepancies between models and makes appear with time, and it is then when the different manufacturing qualities and processes will show up. The procedure explained in this chapter can also be applied to used meters by only changing the specification limits and AQL.

10

Standards and technical guidelines

10.1. INTRODUCTION

Standards and technical guidelines are fundamental when dealing with water meters, for they are closely related to regulations and requirements that the utilities must follow. It is however, difficult to present a chapter on standards and technical guidelines that will remain updated for a long period of time. These documents are of a changing nature, and consequently this part of the book can only be a snapshot of the current situation that will obviously need updating in the future.

However, the fundamentals behind all standards and guidelines are essentially the same. And although the definitions may differ slightly, and it is not easy task to compare documents of different nature and origin, this chapter intends to provide a general overview of the most relevant technical guidelines and standards regarding water meters: from the general principles to the differences between documents and their comparison when possible.

Regardless of the characteristics of each meter type, there are some parameters which are common to all of them. These parameters allow characterizing the different aspects of a meter's performance. The so-called *metrological* parameters are aimed to assess the uncertainty in the measure and the error that can be expected from a certain meter. The *technical* parameters relate to the operation of the meter and its installation.

10.2. METROLOGICAL CHARACTERISTICS

The following parameters are directly related with the primary function of a meter, that is to adequately measure the volume of water that has circulated through it:

- *Actual volume*: Volume of water circulated through the meter in a predefined lapse of time, regardless of its duration. The *actual flowrate* is then defined as the ratio between the actual volume and the predefined time.
- *Indicated volume*: Reading of the meter corresponding to the actual volume. The *indicated flowrate* is defined as the ratio between the indicated volume and the predefined time.
- *Error of indication*: Difference between the indicated and the actual volume. This volume difference may be both positive and negative.
- *Relative error*: Defined as the ratio between the error of indication and the actual volume. It is a non-dimensional parameter that is usually expressed in percentage. Likewise, the relative error may also be positive or negative.
- *Error curve*: The relative error of a meter is not constant or independent of the operating conditions, and changes with flowrate. The error curve is the graphical representation of the evolution of the relative error with flowrate.

Generally speaking, the error curve of any meter presents three zones. In the first one the error is -100%, which means that none of the volume is registered and the meter is stopped. The second zone represents the flowrates that actually start the meters' sensor, although the water consumption is registered with significant errors. The third zone corresponds to values of the flowrate which deliver a lower and reasonably constant error which is almost not dependent on the flowrate.

Figure 10.1 shows the characteristic shape of a water meter's error curve. Considering this curve in detail, the following characteristics can be observed:

- The curve does not reach the left axis. This implies that there is a threshold flowrate for which the sensor of the meter (turbine, piston, disc, etc.) does not move. Values of the flowrate lower than this threshold are not registered by the meter. Consequently, the curve is only represented from a certain value of the flowrate onwards. This value is called the *startup flowrate*. In practice, the startup flowrate cannot be expressed in a single value. It should be considered as the range of flowrates in which, depending on the circumstances, the water transmits enough *energy* to the sensor element of the device to start registering volumes.
- Once the startup flowrate is reached the curve increases its value steeply until the error reaches a value close to zero. In the range of flowrates corresponding to this zone, the error values are negative starting approximately at -70%. It is quite usual for the amplitude of this range to be smaller than 10–15 l/h. In other words, for a new meter the flowrate needed to produce an error close to zero is only 10–15 l/h higher than the startup flowrate.
- Finally, there is a third segment in which the error curve practically remains constant with the flowrate and the error is kept within the admissible error tolerances. The form adopted by the curve in this segment depends on the constructive characteristics of the meter and its technology. The normal operating flowrates of the meter belong to this section.

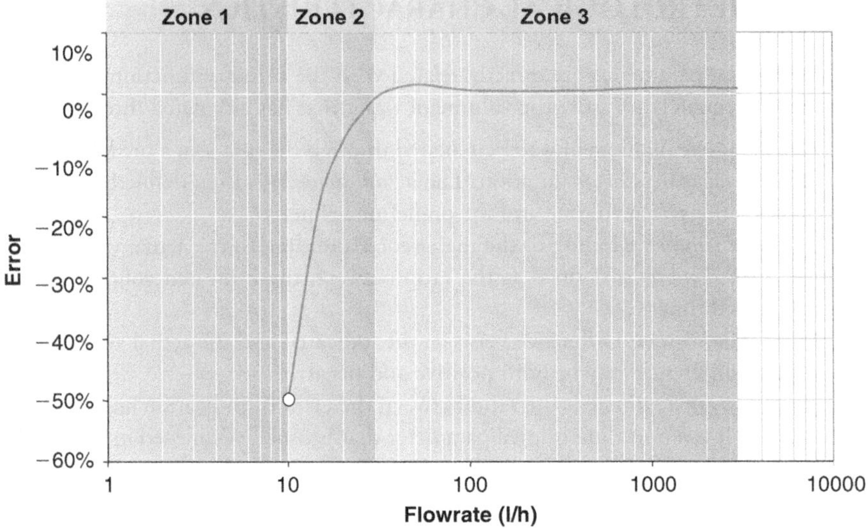

FIGURE 10.1.   ERROR CURVE OF A WATER METER

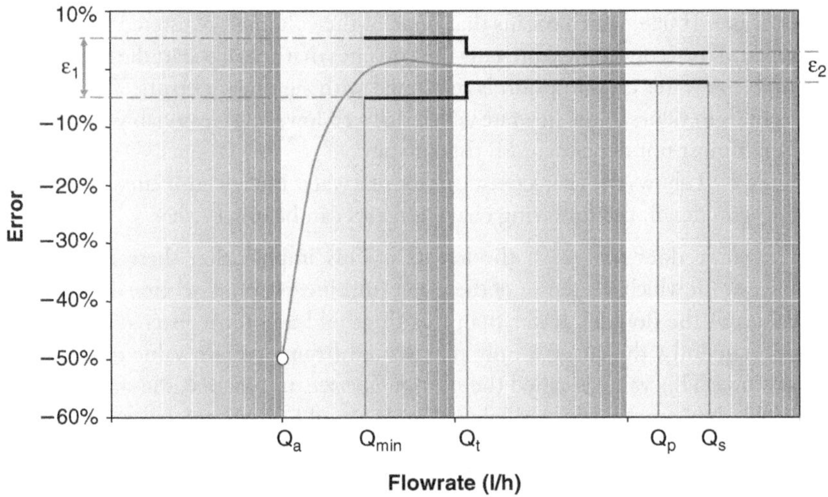

FIGURE 10.2.   MAXIMUM PERMISSIBLE ERRORS OF A WATER METER

A detailed knowledge of the error curve of a meter is critical for any evaluation of its reliability. For this matter, the different international standards pay special attention to the requirements for this curve. The ISO 4064, EN 14154 and the OIML R49 recommendation divide the range of flowrates in two bands, each with a different maximum allowable error. Figure 10.2 shows both zones:

1. The lower zone, defined between the minimum flowrate (Q_m) and the transitional flowrate (Q_t), has a maximum allowable error, ε_1 of 5%.
2. The upper zone, ranging from the transitional flowrate (Q_t) to the overload or maximum (Q_s) flowrate, has a more restrictive error value ε_2 of 2%.

The error curve has great importance. To a great extent, the quality of a meter is the quality of its error curve and its evolution with time. This quality can be defined through the following metrological parameters:

- *Startup flowrate (Q_a)*: A parameter of difficult definition, it does not appear in any standard published to date. It can be defined as the value of the flowrate that generates motion in the meter when the mechanism is at rest. The same concept could be expressed as the *stoppage flowrate* meaning the minimum flowrate necessary to maintain the meter in motion. This definition assesses more accurately the capacity of a meter to measure a leak.

  The startup flowrate (or stoppage flowrate) is useful in determining the percentage of volume registered by a meter, for it is the value of the flowrate in which the error ceases to be -100%. It is therefore the leftmost point of the error curve.
- *Minimum flowrate (Q_1)*: The lowest flowrate at which the meter is required to operate within the maximum permissible error, ε_1.
- *Transitional flowrate (Q_2)*: Flowrate situated between the minimum flowrate and the nominal (ISO 4064:1993) or permanent (ISO 4064:2005, EN 14154, OIML R49:2003) flowrate. It divides the curve into two zones, the lower zone and the upper zone. Each of the zones is characterized by its own maximum allowable error.
- *Permanent flowrate (Q_3)*: The highest flowrate within the nominal working conditions. The meter is required to operate in a satisfactory manner within the maximum allowable error. This flowrate is also known as nominal flowrate (ISO 4064:1993).
- *Overload flowrate (Q_s)*: Value of the flowrate at which the meter is required to operate for short periods of time while maintaining the measuring error within the allowable limits and recovering full operating performance when returning to normal conditions. Sometimes this value is referred to as maximum flowrate (ISO 4064:1993).

The parameters introduced up to this point define the metrological capacity of a water meter. In any case, given the high number of circumstances in which water meters are needed, there should be consideration for different degrees of quality in their performance for each one of them.

These different requirements for a meter's performance (determined by the operating conditions) are covered by the standards by means of *metrological classes*. Meters with equal capacity, that is a same permanent flowrate, will have different minimum and transitional flowrates depending on the metrological class (generally speaking, the higher the quality of the class, the lower the values of the flowrates). For instance, ISO 4064:1993 divides the meters into four qualities or metrological classes: Classes A, B, C and D. The ISO 4064:2005, OIML R49:2003 and EN 14154:2004 Standards define a much wider range of classes. In any case, it is important to point out that the maximum allowed errors in all cases are the same, and that the only thing that changes from a metrological class to the next one are the values of the flowrates at which the meter is required to conform to such values.

10.3. TECHNICAL PARAMETERS

These parameters are referred to technical aspects of the meter, and also to the conditions in which the meter operates. The main ones are:

Pressure

- *Working pressure*: Average recommended pressure for the normal operation of the meter. The operating pressure can range from the *minimum admissible working pressure* to the *maximum admissible working pressure*, which are the limit values that guarantee that the meter will retain its metrological characteristics.
- *Upper limit pressure*: Highest pressure that the meter can withstand for a short period of time without deteriorating.

Working temperature

Average water temperature between the meter entry and exit. The limit values (*maximum* and *minimum admissible working temperatures*) are the ones that the meter can withstand permanently without its metrological characteristics deteriorating.

Pressure loss

This is the head loss caused by the meter. In normal operating conditions, this magnitude must be lower than the *maximum pressure loss within rated operating conditions*.

Operating conditions

Standards often group the temperature and pressure values under what is known as the meter operating conditions. Depending on the level of requirement for the meter they can be grouped under three different levels of operating conditions:

- *Rated operating conditions*: Conditions under which the measuring errors must remain below the maximum admissible values.
- *Reference conditions*: Conditions specified for meter testing or for comparing measurement results.
- *Limiting conditions*: Extreme conditions that the meter should withstand temporarily without suffering any residual damage or deterioration in the measuring error once it returns to normal operating conditions.

Flow direction

- *Forward flow*: Proper direction of the flow according to the design for a normal operation of the meter.
- *Reverse flow*: Direction of the flow opposed to forward flow. The behaviour of a water meter with a reverse flow may vary depending on the standard. Under reverse flow conditions, a meter may:
  - Maintain the last reading registered with forward flow.
  - Register the amount of water circulated in reverse direction independently from the volume circulated in the forward direction.
  - Subtract the registered volume in reverse to the forward value.

Indicating device

This is the element of the meter in charge of displaying the registered values. Its main characteristics as described by the standards are:

- *Measuring units* in which the registered volume is shown. Usually the totalized volume is displayed in multiples of cubic meter.
- *Maximum reading capacity*: Largest figure that the indicating device can display. It should be high enough to display the volume registered after a certain number of hours at the nominal flowrate.
- *Colour coding* for the multiples and submultiples of the measuring units in which the registered volume is expressed. Usually black and red, respectively.
- *Resolution of the indicating device*: Minimum volume which is possible to discern in the indicating device. The resolution is a parameter to be taken into account during the test, in the laboratory or in the field, of water meters.

Designation, dimensions and design

Three final items which are covered in the standards are also important for the selection and management of water meters. The designation of the meter must clearly appear, and is usually based on its diameter or its permanent flowrate. All aspects regarding dimensions and design (such as the diameter, length, threads, housing, materials, freezing protection, etc.) are covered in the standards in different degrees of detail. The most relevant aspects are covered later in the chapter.

10.4.   ISO STANDARDS

The year 2005 saw the publication of the new revision of the ISO Standard for water meters. More precisely the ISO 4064:2005 Standard for *Measuring of water flow in fully charged closed conduits – Meters for cold and hot potable water*. The new standard presents significant differences compared with the previous version, dating from 1993. However, and due to the fact that the 1993 standard applies to meter models approved before October 2006, and since the validity of the model approval lasts for 10 years, it is more than likely that there will be meter models conforming to this standard even in the years to come.

ISO 4064:1993. Measurement of water flow in close conduits – meters for cold potable water.

Issued by: International Organization for Standardization (1993).

Nature: International Standard.

Scope: Cold water meters (up to 30°C) with permanent flowrate values ranging from 0.6 to 4000 m^3/h, and maximum admissible working pressures equal or higher than 10 bar.

The ISO 4064:1993 Standard is structured in three separate parts:

1. Specifications.
2. Installation requirements.
3. Test methods and equipment.

Some of the key issues particular to this standard are:

- The 1993 version is only applicable to integrating measuring devices that determine continuously the volume of water flowing through them, excluding any other liquid. The sensor element must be mechanical and can be formed by volumetric chambers of moving walls (such an oscillating piston) or by any moving element that reacts to the velocity of the circulating water (e.g. single jet meter).
- The designation of the meter is determined by the value of the permanent flowrate, Q_p, expressed in cubic meters per hour and preceded by the letter "N". Such parameter is limited depending on the type of meter to the following values:
  - Threaded connection meters: 0.6, 1, 1.5, 2.5, 3.5, 6, 10.
  - Flanged connection (volumetric, single and multiple jet meters): 15, 20, 30, 50.
  - Flanged connection (Woltmann meters): 15, 25, 40, 60, 100, 150, 250, 400, 600, 1000, 1500, 2500, 4000.
- Four metrological classes are specified: A, B, C and D. Table 10.1 shows their characteristics depending on the permanent flowrate (it must be noted that the ISO 4064:1993 makes reference to the N designation of the meter. Table 10.1 refers them to the permanent flowrate for an increased clarity).
- The value of the overload flowrate, Q_s, is obtained for all metrological classes as twice the permanent flowrate, Q_p.
- The values of the minimum, transitional and maximum flowrates for all the metrological classes are displayed in Table 10.2.
- The maximum admissible error for each flowrate range is $\pm 5\%$ for ε_1 and $\pm 2\%$ for ε_2. Figure 10.3 shows the maximum measuring error for all classes corresponding to a Q_p flowrate of $1.5\,\mathrm{m^3/h}$. The more restrictive nature of Classes C and D can be seen affecting only the values of flowrates at which the meters are required to present errors below the maximum admissible ones.
- The minimum admissible working pressure for a meter is 10 bar for all meters, independently of their type, diameter and metrological class. When the admissible working pressure is higher than 10 bar, the actual value will be displayed visibly in the exterior of the meter.

TABLE 10.1. METROLOGICAL CLASSES AS DEFINED IN ISO 4064:1993

Class	Flowrate	Q_p	
		$<15\,\mathrm{m^3/h}$	$\geqslant 15\,\mathrm{m^3/h}$
A	Q_m	$0.04Q_p$	$0.08Q_p$
	Q_t	$0.10Q_p$	$0.30Q_p$
B	Q_m	$0.02Q_p$	$0.03Q_p$
	Q_t	$0.08Q_p$	$0.20Q_p$
C	Q_m	$0.01Q_p$	$0.006Q_p$
	Q_t	$0.015Q_p$	$0.015Q_p$
D	Q_m	$0.0075Q_p$	–
	Q_t	$0.0115Q_p$	–

TABLE 10.2. CHARACTERISTIC FLOWRATES PER METROLOGICAL CLASS IN ISO 4064:1993

Q_p (m³/h)	Class A		Class B		Class C		Class D		Q_s (l/h)
	Q_m (l/h)	Q_t (l/h)	Q_m (l/h)	Q_t (l/h)	Q_m (l/h)	Q_t (l/h)	Q_m (l/h)	Q_t (l/h)	
0.6	24	60	12	48	6	9	4.50	6.90	1.2
1	40	100	20	80	10	15	7.50	11.50	2
1.5	60	150	30	120	15	22.5	11.25	17.25	3
2.5	100	250	50	200	25	37.5	18.75	28.75	5
3.5	140	350	70	280	35	52.5	26.25	40.25	7
6	240	600	120	480	60	90	45.00	69.00	12
10	400	1000	200	800	100	150	75.00	115.00	20
15	1200	4500	450	3000	90	225	–	–	30
20	1600	6000	600	4000	120	300	–	–	40
25	2000	7500	750	5000	150	375	–	–	50
30	2400	9000	900	6000	180	450	–	–	60
40	3200	12000	1200	8000	240	600	–	–	80
50	4000	15000	1500	10000	300	750	–	–	100
60	4800	18000	1800	12000	360	900	–	–	120
100	8000	30000	3000	20000	600	1500	–	–	200
150	12000	45000	4500	30000	900	2250	–	–	300
250	20000	75000	7500	50000	1500	3750	–	–	500
400	32000	120000	12000	80000	2400	6000	–	–	800
600	48000	180000	18000	120000	3600	9000	–	–	1200
1000	80000	300000	30000	200000	6000	15000	–	–	2000
1500	12000	450000	45000	300000	9000	22500	–	–	3000
2500	200000	750000	75000	500000	15000	37500	–	–	5000
4000	320000	1200000	120000	800000	24000	60000	–	–	8000

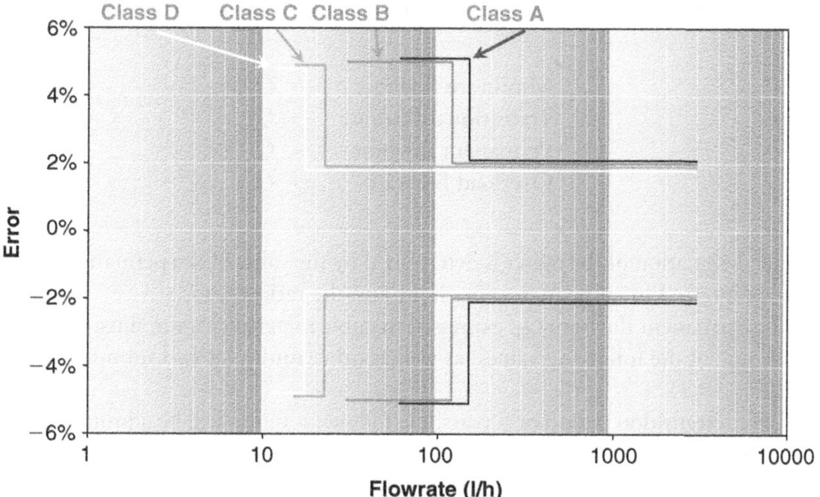

FIGURE 10.3. ERROR LIMITS FOR DOMESTIC METERS WITH Q_p 1.5 M³/H – ISO 4064:1993

- The maximum admissible working temperature is set to 30°C for all cold water meters.
- The maximum pressure loss in the meter, in bars, must be restricted to one of the following values: 0.1, 0.3, 0.6 and 1 bar.

ISO 4064:2005. Measurement of water flow in close conduits – meters for cold and hot potable water.

Issued by: International Organization for Standardization (2005).

Nature: International Standard.

Scope: Cold and hot water meters with maximum admissible operating pressures equal or higher to 1 MPa, or 0.6 MPa for meters with diameters equal or larger than 500 mm and maximum temperatures of 30°C for cold water and 180°C for hot water, depending on the class.

The ISO 4064:2005 Standard is divided in three sections:

1. Specifications.
2. Installation requirements.
3. Test methods and equipment.

Particular to this version:

- This standard is also applicable to any water meter that incorporates any electrical or electronic principles and mechanical meters with electronic components. As a matter of fact, the standard is applicable to any water meter, defined as a measuring device that measures continuously the water volume circulated through it.
- The meters are characterized by the nominal diameter of the connections or the flanges. For each diameter, several dimensions of the meter are provided for length, height and width.
- The flowrates characterizing the meter's metrology can be defined as follows:

$$
\begin{array}{lcl}
\text{Minimum flowrate} & \rightarrow & Q_1 \\
\text{Transitional flowrate} & \rightarrow & Q_2 \\
\text{Permanent flowrate} & \rightarrow & Q_3 \\
\text{Overload flowrate} & \rightarrow & Q_4
\end{array}
$$

- The designation of the meter is determined by the value of the permanent flowrate, Q_3, expressed in cubic meters per hour and the ratio Q_3/Q_1.
- The permanent flowrate Q_3, expressed in cubic meters per hour, must be restricted to some of the following values, as well as other multiples and submultiples of 10 (Table 10.3).

  For a transition period of 5 years, the following values will be admitted: 1.5, 3.5, 6, 15 and 20 m^3/h.
- The minimum flowrate, and as a consequence, the metrological quality of the meter, is defined by the ratio Q_3/Q_1. The admissible values for that ratio are shown in Table 10.4. Also accepted are other multiples and submultiples of 10.

TABLE 10.3. ADMISSIBLE VALUES FOR PERMANENT FLOWRATE

Q_3				
1	1.6	2.5	4	6.3
10	16	25	40	63
100	160	250	400	630
1000	1600	2500	4000	6300

TABLE 10.4. ADMISSIBLE VALUES FOR Q_3/Q_1 RATIO

Q_3/Q_1									
10	12.5	16	20	25	31.5	40	50	63	80
100	125	160	200	250	315	400	500	630	800

For a transition period of 5 years, the following values will be admitted: 15, 35, 60 and 212.

- The value of the Q_3/Q_1 ratio may be preceded by the letter R. A meter with a ratio of 200 may be denominated as R200. The metrological classes are not explicitly defined in this standard, and are rather the result of the different values of the R ratio (Q_3/Q_1), for the other flowrates (Q_2 and Q_4) are defined from the values of Q_1 and Q_3.
- The transitional flowrate is established from the ratio Q_2/Q_1, with a single value of 1.6. However, and for a 5 year transition period the following values will be accepted: 1.5, 2.5, 4 and 6.3.
- The overload flowrate, Q_4, is defined from the ratio Q_4/Q_3, with a single value of 1.25.
- The maximum admissible error for both flowrate ranges, determined by the transitional flowrate, are established as follows:

$$\varepsilon_1 = \pm 5\%$$

$$\varepsilon_2 = \pm 2\% \text{ for } \quad T \leqslant 30°C$$

$$\varepsilon_2 = \pm 3\% \text{ for } \quad T > 30°C$$

Figure 10.4 shows the error limits for cold water meters with a Q_3 equal to 2.5 m^3/h according to the ISO 4064:2005 Standard.

- The minimum admissible working pressure is fixed and equal to 0.3 bar for all meters. The maximum operating pressure is used to group the meters in the classes shown in Table 10.5.
- The maximum and minimum operating temperatures group the meters in the classes shown in Table 10.6.

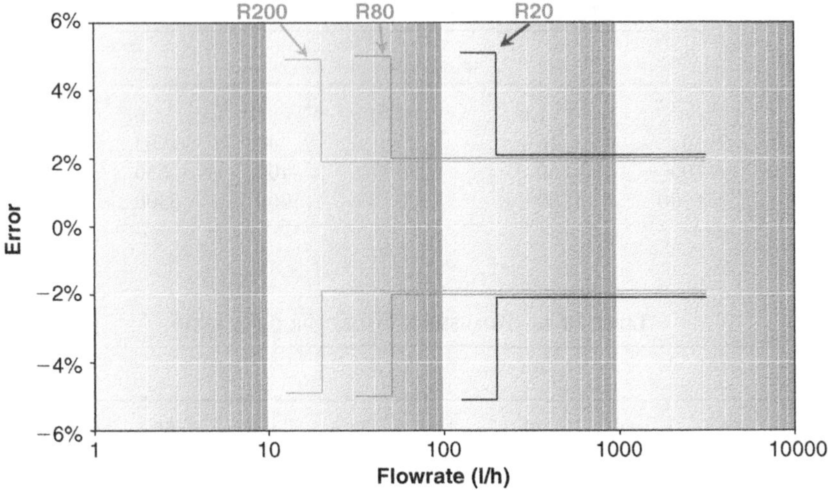

FIGURE 10.4. ERROR LIMITS FOR COLD WATER DOMESTIC METERS WITH Q_p 2.5 M³/H (ISO 4064:2005)

TABLE 10.5. METER CLASSES ACCORDING TO MAXIMUM OPERATING PRESSURE ISO 4064:2005

Class	Maximum operating pressure (bar)
MAP 6	6
MAP 10	10
MAP 16	16
MAP 25	25
MAP 40	40

TABLE 10.6. METER CLASSES ACCORDING TO OPERATING TEMPERATURES ISO 4064:2005

Class	Operating temperatures (°C)	
	Minimum	Maximum
T30	0.1	30
T50	0.1	50
T70	0.1	70
T90	0.1	90
T130	0.1	130
T180	0.1	180
T30/70	30	70
T30/90	30	90
T30/130	30	130
T30/180	30	180

- The maximum pressure loss that the meter can create in normal operating conditions with a permanent flowrate, Q_3, is 0.63 bar. Additional values below this one define the classes corresponding to the ones shown in Table 10.7.
- One of the main novelties in the ISO 4064:2005 is the classification of the devices as a function of their sensitivity to distortions in the velocity profile. Meters are classified according to the length of straight pipe needed, upstream and downstream, to conform to the maximum admissible errors. This is defined in Table 10.8 for the installation upstream from the meter and in Table 10.9 for the installation downstream from the meter. For each class, the tables show the length of pipe that the meter needs at each side, expressed as a multiple of its nominal diameter, and the possible requirement of a flow tranquilizer to guarantee that the error is maintained within the required limits ε_1 and ε_2.

TABLE 10.7. METER CLASSES ACCORDING TO THE MAXIMUM PRESSURE LOSS

Class	Maximum pressure loss (bar)
ΔP 63	0.63
ΔP 40	0.40
ΔP 25	0.25
ΔP 16	0.16
ΔP 10	0.10

TABLE 10.8. METER CLASSES ACCORDING TO THE SENSITIVITY OF THE METER TO IRREGULARITIES IN THE UPSTREAM VELOCITY PROFILE ISO 4064:2005

Class	Length of pipe needed ($\times$DN)	Need for a tranquilizer
U0	0	No
U3	3	No
U5	5	No
U10	10	No
U15	15	No
U0S	0	Yes
U3S	3	Yes
U5S	5	Yes
U10S	10	Yes

TABLE 10.9. METER CLASSES ACCORDING TO THE SENSITIVITY OF THE METER TO IRREGULARITIES IN THE DOWNSTREAM VELOCITY PROFILE ISO 4064:2005

Class	Length of pipe needed ($\times$DN)	Need for a tranquilizer
D0	0	No
D3	3	No
D5	5	No
D0S	0	Yes
D3S	3	Yes

TABLE 10.10. MINIMUM READING RANGES FOR THE
INDICATING DEVICE ISO 4064:2005

Q_p (m³/h)	Minimum reading range (m³)
$Q_p \leq 6.3$	9999
$6.3 < Q_p \leq 63$	99999
$63 < Q_p \leq 630$	999999
$630 < Q_p \leq 6300$	9999999

- The climatic and mechanical conditions in the meter environment are taken into account defining three classes to this respect:
  - *Class B*: Fixed outside meters.
  - *Class C*: Fixed inside meters.
  - *Class I*: Mobile meters.

  For each one of these classes a certain severity level is set regarding dry heat, humid heat (and cyclic), cold, vibrations and impacts, which should be considered in the corresponding laboratory tests.
- The electromagnetic conditions present in the vicinity of the meter are defined in two specific classes:
  - *Class E1*: Domestic, commercial and light industry environment.
  - *Class E2*: Heavy industry environment.

  For each one of these classes a certain severity level is defined regarding electrostatic discharges, radio frequency emission fields, overloads and transients/electrical failures, which should be considered in the corresponding laboratory tests.
- The indicating device of each meter will have a minimum reading range, in cubic meters, that is fixed as a function of Q_p, as shown in Table 10.10.

10.5. RECOMMENDATIONS OF THE INTERNATIONAL ORGANIZATION OF LEGAL METROLOGY

Similarly to ISO Standards, there is a new version of recommendation 49 by the International Organization of Legal Metrology (OIML) that begun its development in 2000 and was published in 2003. This new version is very similar to the ISO 4064:2005 and also to the European Standard EN 14154.

OIML R49:2003. Water meters intended for the metering of cold potable water.

Issued by: International Organization of Legal Metrology (2003).

Nature: Recommendation that all meters need to fulfil in those countries in which these devices are subject to control by the administration.

Scope: Meters with indicating device for the measured volume, for cold drinking water in pressurized pipe.

TABLE 10.11. METROLOGICAL CLASSES IN OIML R49:2003

		$Q_p < 100\,\text{m}^3/\text{h}$		$Q_p \geqslant 100\,\text{m}^3/\text{h}$	
		$0.3°\text{C} \leqslant T \leqslant 30°\text{C}$	$T > 30°\text{C}$	$0.3°\text{C} \leqslant T \leqslant 30°\text{C}$	$T > 30°\text{C}$
Class 1	ε_1	–	–	3%	3%
	ε_2	–	–	1%	3%
Class 2	ε_1	5%	5%	5%	5%
	ε_2	2%	3%	2%	3%

The OIML R49 is divided in three parts:

1. Technical and metrological requirements.
2. Testing methods.
3. Test report format.

It is important to point out that the OIML recommendation and the ISO 4064:2005 are practically identical, except for small details. As a consequence, only the relevant differences will be presented here:

- The new OIML recommendations, similarly to other new standards, do not specify the physical principle used to measure the volume. There are consequently no restrictions to the type of sensor used, which may be mechanical, electrical or electronic. The only specification is that a meter must have a sensor element, a calculator and an indicating device.
- Similarly to the ISO 4064:2005 the metrological characteristics of the meter are defined with the values of Q_1, Q_2, Q_3 and Q_4. The admissible values for Q_3 and the ratios Q_3/Q_1, Q_2/Q_1 and Q_4/Q_3 are the same than the ones presented in the ISO 2005 Standard.
- However, contrary to the ISO document, two meter classes are defined according to the maximum admissible error values, as a function of Q_3 and the temperature of water. Table 10.11 shows these classes.
- The pressure loss, including the filter that the meter may have, should not exceed 0.1 MPa (1 bar) between Q_1 and Q_4. This recommendation does not classify meters according to the generated pressure loss.
- The inscriptions that must be displayed in the meter's housing are:
  - Measuring unit (m^3).
  - Metrological class if it is different than 2.
  - Numerical value of Q_3.
  - Numerical value of the ratio Q_3/Q_1, preceded by the letter "R".
  - Numerical value of the ratio Q_2/Q_1, if different from 1.6.
  - Symbol of model approval in accordance with local regulations.
  - Name or logo of the manufacturer.
  - Manufacturing year (two last digits) and serial number.
  - Regular flow direction.
  - Maximum operating pressure when above 0.1 MPa (10 bar).

TABLE 10.12. MINIMUM READING RANGES FOR THE INDICATING DEVICE, **OIML R49:2003**

Q_3 (m^3/h)	Minimum reading range (m^3)
$Q_3 \leq 6.3$	9999
$6.3 < Q_3 \leq 63$	99999
$63 < Q_3 \leq 630$	9999999
$630 < Q_3 \leq 6300$	99999999

- Letter "V" or "H" depending on whether the meter must be installed vertically or horizontally.
- Temperature class, if different from T30 (Table 10.6).
- The manufacturer may display the maximum pressure loss.

For electronic meters:
- For external electrical supply: voltage and frequency.
- If replaceable batteries are used, date of replacement.
- If non-replaceable batteries are used, meter replacement date.
- The range of the device must follow the restrictions included in Table 10.12.
- The cubic meters and their multiples will be displayed in black. The submultiples of cubic meters will be displayed in red.

10.6. STANDARDS FROM THE EUROPEAN COMMITTEE FOR STANDARDIZATION (CEN)

There are currently two European Standards published by European Committee for Standardization (CEN). The first one refers to clean drinking water (EN 14154:2005) and the other one for irrigation meters (EN 14268:2005).

CEN EN 14154.

Issued by: European Committee for Standardization (2005).

Nature: European Standard.

Scope: Water meters intended for residential, commercial, light industrial and industrial use to meter the actual volume flow of cold and heated water.

The EN 14154 Standard has three parts:

14154-1: General requirements.
14154-2: Installations and use conditions.
14154-3: Test methods and equipment.

Also, the CEN Standard is almost identical to the ISO 4064:2005, and the information previously presented therefore applies.

CEN EN 14268. Irrigation techniques – meters for irrigation water.

Issued by: European Committee for Standardization (2005).

Nature: European Standard.

Scope: Meters fitted to irrigation machines and networks used to measure the actual volume of irrigation water flowing through a fully charged close conduit.

Although apparently this standard might not be of great interest to water utility managers, the truth is that any meter that is not used for clean water would be out of the scope of the EN 14154 and the new 2004/22/EC Directive. Under these circumstances the EN 14268 becomes relevant for many applications in the water industry in which the water is not clean or is used for industrial purposes.

Differentiating aspects in this standard:

- The meter designation is determined by the value of Q_3, expressed in cubic meters per hour, and the ratio Q_3/Q_1. The proposed values are the same as in the EN 14154:2005 and the ISO 4064:2005.
- Two metrological classes are defined according to the Q_3/Q_1 ratio:
  - Class A for $Q_3/Q_1 \geqslant 10$.
  - Class B for $Q_3/Q_1 \geqslant 25$.
- It does not contemplate the existence of Q_2, and consequently the maximum measuring error will be the same between the minimum flowrate, Q_1, and the overload flowrate, Q_4.
- Q_4 is defined from the ratio Q_4/Q_3, with a single value of 1.25.
- The maximum admissible error for a single range of flowrates is set to $\pm 5\%$. Figure 10.5 shows the error limits for meters with Q_3 equal to 63 m³/h, reflecting the consequences of not having Q_2.

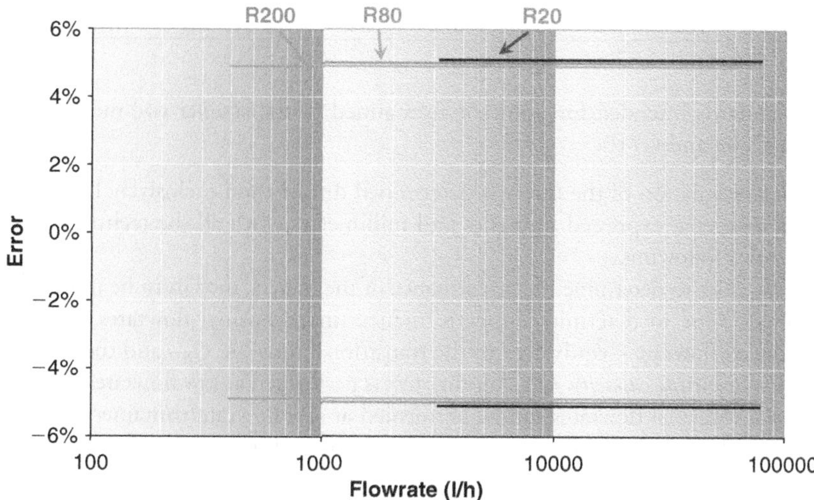

FIGURE 10.5.   ERROR LIMITS FOR METERS WITH $Q_3 = 63$ M³/H EN 14268

- The maximum admissible working pressure is implicitly set at 10 bar.
- The range of admissible water temperatures is from 30°C to 0.1°C.
- The maximum pressure loss that the meter can produce at a permanent flowrate is 0.065 MPa (0.65 bar).
- Especial attention is paid (almost half the standard) to the tests needed for the model approval. Worthy of mention are the tests to check the sturdiness against dirty waters and the wear due to normal use.

10.7. AMERICAN WATER WORKS ASSOCIATION/ AMERICAN NATIONAL STANDARD INSTITUTE STANDARDS

In the United States there are no standards of obligatory compliance like in the European Union. However, American Water Works Association (AWWA) and American National Standard Institute (ANSI) publish several American National Standards (AWWA Standards) regarding water meters. These standards are of a voluntary nature, although manufacturers are encouraged to apply them and publicize that their products are in accordance with such documents. Just like any other standards, the documents are elaborated by consensus among the main interested stakeholders.

ANSI/AWWA C700 to C750.

Issued by: American National Standard Institute (ANSI) and American Water Works Association (AWWA) (1996–2002).

Nature: American National Standard.

Scope: Cold water meters. Further details on the scope of each document are included in Table 10.13.

The standards intended for domestic uses aimed to the smaller size meters are the C700, C708 and C710:

- The designation of the meter is determined directly and exclusively by its nominal diameter, expressed in inches and millimetres, with all the technical characteristics following.

   In order to determine the performance of the meters, they must be tested in two ranges. One to determine its performance under normal flowrates (between a reduced flowrate – equivalent to the transitional flowrate Q_t – and the *safe maximum operating capacity*, Q_s). Another test is performed at low flowrates. While the test for normal flowrates can be performed at any flowrate contained within the range, the low-flowrate test must necessarily be performed at a specific value designated as *minimum test flow*, Q_m.

   The values of the flowrates for the tests are specifically associated to each nominal diameter. The relationship between the flowrate and the diameter is shown

TABLE 10.13.   ANSI/AWWA STANDARDS FOR COLD WATER METERS

Standard	Characteristics of the meter	Use	Diameter range
C700-02	Displacement type (bronze main case)	Customer service	0.5″ (13 mm) 2″ (50 mm)
C701-02	Turbine type	Customer service	1.5″ (40 mm) 12″ (320 mm)
C701-01	Compound type		2″ (50 mm) 10″ (250 mm)
C703-96	Cold water meters	Fire service	3″ (75 mm) 10″ (250 mm)
C704-02	Propeller type	Waterworks applications	2″ (50 mm) 72″ (1800 mm)
C706-96 (R01)	Direct-reading, remote-registration systems	Customer service	
C707-82 (R92)	Encoder-type remote-registration systems	Customer service	
C708-96	Multiple jet	Customer service	0.625″ (16 mm) 2″ (50 mm)
C710-02	Displacement type (plastic main case)	Customer service	0.5″ (13 mm) 1″ (25 mm)
C712-02	Single jet	Customer service	1.5″ (40 mm) 6″ (150 mm)
C750-03	Transit-time flowmeters	Waterworks applications	

in Table 10.14. Additionally, a normal operating flowrate is defined equivalent to the *permanent flowrate*, Q_p, that is usually about half of the safe maximum operating capacity, Q_s. A summary of the flowrate values for the different diameters and standards is shown in Table 10.14.

- Compared to other international standards, the ANSI/AWWA Standards differ in specifying different errors depending on the meter type. For instance, the errors for multiple jet meters are maintained symmetrical around the zero value, while this is not the case for volumetric meters. Table 10.15 and Figure 10.6 compare numerically and graphically the requirements for the different meter types.

  Logically, ANSI/AWWA Standards do not define metrological classes within a meter denomination type. Consequently, and for each nominal diameter:
  - The values of Q_m and Q_t are always the same, regardless of any other consideration.
  - The variations in the error limits are only found between different meter types, and as a consequence, defined in different standards.

- The minimum admissible working pressure (so the meter can operate continuously without leaks or damage) is set for every diameter at 10.5 bar (150 psi = 1050 kPa).

TABLE 10.14. CHARACTERISTIC FLOWRATES IN THE ANSI/AWWA STANDARDS

Standard	DN (in.)	DN (mm)	Q_m	Q_t	Q_p	Q_s
C700	½	13			7.5 gpm	15 gpm
C710	½ × ¾	13 × 19	¼ gpm	1 gpm	1.7 m³/h	3.4 m³/h
	5/8	16	0.06 m³/h	0.2 m³/h	10 gpm	20 gpm
C700	5/8 × ¾	16 × 19			2.3 m³/h	4.5 m³/h
C708	¾	19–20	½ gpm	2 gpm	15 gpm	30 gpm
C710			0.110 m³/h	0.5 m³/h	3.4 m³/h	6.8 m³/h
	1	25	¾ gpm	3 gpm	25 gpm	50 gpm
			0.170 m³/h	0.7 m³/h	5.7 m³/h	11.4 m³/h
C700	1½	38–40	1½ gpm	5 gpm	50 gpm	100 gpm
C708			0.340 m³/h	1.1 m³/h	11.3 m³/h	22.7 m³/h
	2	50–51	2 gpm	8 gpm	80 gpm	160 gpm
			0.450 m³/h	1.8 m³/h	18.2 m³/h	36.3 m³/h
C712	1½	40	½ gpm	1.5 gpm	50 gpm	100 gpm
			0.110 m³/h	0.34 m³/h	11 m³/h	23 m³/h
	2	50	½ gpm	2 gpm	80 gpm	160 gpm
			0.110 m³/h	0.45 m³/h	18 m³/h	36 m³/h
	3	80	½ gpm	2.5 gpm	160 gpm	320 gpm
			0.110 m³/h	0.57 m³/h	36 m³/h	73 m³/h
	4	100	¾ gpm	3 gpm	250 gpm	500 gpm
			0.170 m³/h	0.68 m³/h	55 m³/h	110 m³/h
	6	150	1 ½ gpm	4 gpm	500 gpm	1000 gpm
			0.340 m³/h	0.91 m³/h	110 m³/h	220 m³/h

TABLE 10.15. MAXIMUM ERRORS ACCORDING TO ANSI/AWWA STANDARDS

	Multiple jet	Turbine Class I	Turbine Class II	Single jet	Volumetric flowrate
Minimum flowrates ε_1	±3%			−5% +1.5%	−5% +1%
Normal flowrates ε_2	±1.5%	±2%	±1.5%	±1.5%	±1.5%

- The maximum admissible working temperature is fixed for every diameter and meter at 27°C.
- The maximum pressure loss is fixed for every diameter and type of meter at 1.03 bar (15 psi = 103 kPa).
- The C700 to C750 Standards pay special attention to the construction materials and their characteristics, (composition, mechanical characteristics, related standards, etc.). Specifications are set for the different pieces and the morphology and design of the meter.

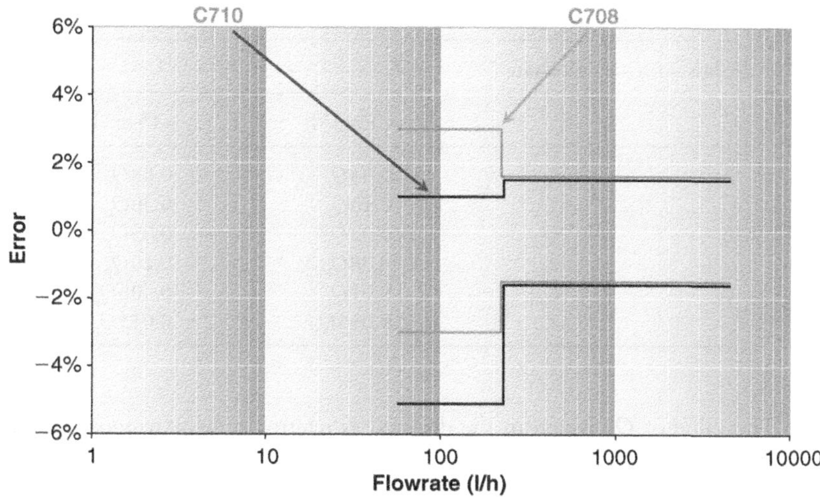

FIGURE 10.6. ERROR LIMITS FOR 5/8″ VOLUMETRIC (C700 AND C710) AND MULTIPLE JET (C708) METERS IN AWWA/ANSI

10.8. EUROPEAN UNION DIRECTIVES

The current applicable European Union Directive on water meters is the 2004/22/EC. This directive replaced two older directives in 2004. However, meter models approved with the old directives will be valid for 10 years after their approval date. For this reason, the requirements of the old directives could be present in models manufactured as far as 2015 (the new directive only became obligatory in late 2006 and the model approval for a water meter has a 10 year validity).

The new directive defines essential requirements that any measuring device must fulfil, and specific requirements for each type of device. The directive also recognizes CEN as the authorized organism to develop harmonized standards that allow assessing the conformity of the devices with the new directive.

Directive 75/33/EEC (derogated by 2004/22/EC).

Issued by: European Union (1974).

Nature: European Union Directive. Obligatory for all models showing the inscription EEC.

Scope: Cold water meters (0–30°C).

The most relevant requirements found in this directive are:

- The designation of the meter is directly determined by the value of Q_p, which in the 75/33/EEC Directive is called nominal, expressed in cubic meters per hour. However, the directive does not set specific values for the nominal flowrate.
- Three metrological Classes, A, B and C, are established in an increasing order of the sensitivity at low flowrates. The definitions for Q_m and Q_t of each metrological class as a function of Q_p are included in Table 10.16.

TABLE 10.16. METROLOGICAL CLASSES IN THE 75/33/EEC DIRECTIVE

Class	Flowrate	Q_p	
		$<15\,\mathrm{m^3/h}$	$\geqslant15\,\mathrm{m^3/h}$
A	Q_m	$0.04Q_p$	$0.08Q_p$
	Q_t	$0.10Q_p$	$0.30Q_p$
B	Q_m	$0.02Q_p$	$0.03Q_p$
	Q_t	$0.08Q_p$	$0.20Q_p$
C	Q_m	$0.01Q_p$	$0.006Q_p$
	Q_t	$0.015Q_p$	$0.015Q_p$

The value of Q_s, which in the directive is referred to as *maximum flowrate*, is defined for all metrological classes as twice the value of Q_p. Table 10.16 shows all values of the characteristic flowrates for every metrological class:

- The maximum admissible error for each flowrate range is set to $\pm5\%$ for ε_1 and $\pm2\%$ for ε_2.
- The maximum admissible working pressure is established, as a minimum value, in 10 bar.
- The maximum pressure loss cannot be greater than 0.25 bar at Q_p, or 1 bar at Q_s. Additionally, the meters are classified into four groups according to the pressure losses at a flowrate Q_s, with values of 1, 0.6, 0.3 and 0.1, respectively.

The directive defines the testing procedures necessary to *approve* each new meter model. The model approval is obligatory to commercialize a new model of meter within the European Union. The testing procedure consists of four successive tests: water-tightness, determination of the error curve at six specific flowrates (between Q_m and Q_s), determination of the pressure loss curve and accelerated endurance test.

The *initial verification* procedure of the meters (applied to every meter sold for custody transfer purposes) comprises tests at three flowrates – Q_m, Q_t and Q_s – to determine the error and the pressure loss (only last flowrate).

Directive 79/830/EEC (derogated by 2004/22/EC).

Issued by: European Union (1979).

Nature: European Union Directive. Obligatory for all models showing the inscription EEC.

Scope: Hot water meters (30–90°C).

The same aspects outlined for the 75/33/EEC Directive are applicable to the 79/830/EEC. The differences are:

- A new metrological Class D is added, and the definition of the flowrates for the other three (A, B and C) changes (Table 10.17).
- The maximum admissible error for each flowrate range is fixed at $\pm5\%$ for ε_1 and $\pm3\%$ for ε_2. As a consequence, the graphical representation of the error limits for

TABLE 10.17. METROLOGICAL CLASSES IN THE 79/830/EEC DIRECTIVE

Class	Flowrate	Q_p	
		$<15\,\mathrm{m^3/h}$	$\geqslant 15\,\mathrm{m^3/h}$
A	Q_m	$0.04Q_p$	$0.08Q_p$
	Q_t	$0.10Q_p$	$0.20Q_p$
B	Q_m	$0.02Q_p$	$0.04Q_p$
	Q_t	$0.08Q_p$	$0.15Q_p$
C	Q_m	$0.01Q_p$	$0.02Q_p$
	Q_t	$0.006Q_p$	$0.010Q_p$
D	Q_m	$0.01Q_p$	
	Q_t	$0.15Q_p$	

all flowrates within the range of the 79/830/EEC will be very similar to Figure 10.3, although not identical.

- For the *model approval* procedure, the 79/830/EEC adds the thermal shock resistance test for meters with a Q_p lower or equal than $10\,\mathrm{m^3/h}$.
- For the *initial verification* procedure, the 79/830/EEC requires the same tests as the 75/33/EEC (but with hot water) and adds an additional watertightness test with cold water at 1.6 times the maximum working pressure for a minute.

Directive 2004/22/EC.

Issued by: European Union (2004).

Nature: European Union Directive. Obligatory for all models showing the inscription EC, starting on 31 October 2006.

Scope: Measuring instruments.

This directive is a general document contemplating any type of instrument subject to legal metrological control. The directive is intended to group all previously existing directives concerning measuring instruments in a single document. Due to its general nature it has two parts. The first part defines the essential requirements (Annex I) that all measuring instruments subject to legal metrological control must fulfil. The second part defines the specific requirements for each type of instrument. More precisely, the specific requirements for water meters are detailed in Annex M1.

The essential requirements defined in the directive are referent to the characteristics necessary to provide a high degree of protection to all parties involved in a commercial transaction resulting from the measurement of a variable. According to the directive, the instruments must be designed in such a way that the user will have a high degree of security regarding the result of the measurement.

There are not many specific requirements for water meters in the new European Directive. The minimum, transitional, permanent and overload flowrates are defined, similarly to the OIML R49:2003 recommendation and the ISO 4064:2005 and EN 14154:2005 Standards.

Regarding the operating range, only the following restrictions are defined:

$Q_3/Q_1 \geqslant 10$
$Q_2/Q_1 = 1.6$
$Q_4/Q_3 = 1.25$ (for a 5 year transition period 1.5, 2.5, 4 or 6.3 are also allowed)

The minimum operating temperature ranges are:

0.1°C and at least 30°C.
30°C and at least 90°C.

The operating pressures range for a meter is 0.3–10 bar.
The maximum admissible errors are:

±2% between Q_2 and Q_4 for water colder than 30°C.
±3% between Q_2 and Q_4 for water hotter than 30°C.
±5% between Q_1 and Q_2 for water at any temperature.

Among the specific requirements, the directive also indicates how the meter should behave under electromagnetic disturbances, and durability and suitability characteristics.

Bibliography

American Water Works Association (1966). Determination of economic period for water meter replacement. *Journal of AWWA*, 58(6); 642–650.

American Water Works Association (1999). *Water Meters – Selection, Installation, Testing and Maintenance*. AWWA Manual M6.

Al-Kahazraji Y.A., Hemp J. (1980). *Electromagnetic Flowmeters and Methods of Measuring Flow*. Patent application No 8011624.

Allender H. (1996). Determining the economical optimum life of residential water meters. *Journal of Water Engineering and Management*, September; 20–24.

Arregui F.J. (1999). *Propuesta de una metodología para el análisis y gestión del parque de contadores de agua*. PhD. Thesis. Universidad Politécnica de Valencia.

Arregui F.J. (2000). *Cálculo del patrón de consumo y la vida útil de los contadores de Santa Fé de Bogotá*. Internal Document. Instituto Tecnológico del Agua. Universidad Politécnica de Valencia.

Arregui F.J. (2002). *Cálculo de las incertidumbres en la estimación del error del parque de contadores*. Internal Document. Instituto Tecnológico del Agua. Universidad Politécnica de Valencia.

Arregui, F.J. (2005). *Determinación del error de medición del abastecimiento de Castellón de la Plana*. Unpublished report.

Arregui F.J., García Serra J., López G., Martínez J. (1998). Metodología para la evaluación del error de medición de un parque de contadores. *Ingeniería del agua*, December; 55–66.

Arregui F.J., Palau C.V., Gascón L., Peris O. (2003). *Evaluating Domestic Water Meter Accuracy. A Case Study*. Pumps, electromechanical devices and systems. Applied to urban water management, Balkema, pp. 343–352.

Baker (1982). *Electromagnetic Flowmeters. Developments in Flow Measurement*. I. Ed. R.W.W. Scott, Chapter 7.

Baker (2000). *Flow Measurement Handbook. Industrial Design, Operating Principles, Performance, and Applications.* Cambridge University Press, Cambridge.

Baker R.C., Deacon J.E. (1983). *Tests on Turbine, Vortex and Electromagnetic Flowmeters in Two Phase Air–Water Upward Flow.* International conference Physical Modelization. Multi-phase Flow. Coventry, England, pp. 33–52.

Bowen P.T., Harp J.F., Hendricks J.E., Shoeleh M. (1991). *Evaluating Residential Meter Performance.* Ed. American Water Works Association Research Foundation AWWARF. Denver, CO.

Bowen P.T., Harp J.F., Baxter J.W., Shull R.D. (1993). *Residential Water Use Patterns.* Ed American Water Works Association Research Foundation. Denver, CO.

Brittain, R.L. (1970). *Small meter periodic test limit extension study.* Unpublished report, Philadelphia Suburban Water Company.

Cascetta F., Palombo A., Scalabrini G. (2003). *Water Flow Measurement in Large Bore Pipes: An Experimental Comparison between Two Different Types of Insertion Flowmeters.* ISA Transactions.

Gary W. (2003). *Gaining Insights on Domestic Water Demand through Remote Sensing.* Center for sustainability of semi-arid hydrology & riparian areas (SAHRA). http://www.sahra.arizona.edu/research/TA5/demand.html

Graeser H.J. (1958). Detecting lost water at Dallas. *Journal of AWWA, Julio;* 925–932.

Hemp J., Sanderson M.L. (1981). *Electromagnetic Flowmeters – A State of the Art Review.* International conference on advances in flow measurement techniques, Warwick, pp 319–340.

Hudson W.D. (1964). Reduction of unaccounted-for water. *Journal of AWWA,* Febrero; 143–148.

Kuranz (1942). Meter maintenance practice. *Journal of AWWA,* 34(1); 117–120.

Kurokawa E., Bornia A.C. (2002). *Uma proposta para a utilizaçao do controle estatístico de processo (CEP) a través da carta X como uma ferramenta gerencial para a avaliaçao da vazao mínima noturna de um setor.* Proceedings Seminario Planejamento, projeto e operação de redes de abastecimento de agua. Joao Pessoa, Brasil.

Male J.W., Noss R.R., Moore I.C. (1985). *Identifying and Reducing Losses in Water Distribution Systems.* Noyes Publications.

Nielsen N. (1969). *Determination of Proper Age for 5/8 Inch Meters in the Hackensack and Spring Valley Systems.* Report for the Hackensack Water Company, Harrington Park, NJ.

Nikuradse J. (1932). *Gesetzmässigkeiten Der Turbulenten Strömung in Glatten Rohren.* Forschg. Arb. Ing. – Wes. No 356.

Palau C.V., Arregui F.J., Ferrer A. (2003). *Using multivariate principal component analysis of injected water flows to detect anomalous behaviours in a water supply system.* Proceedings Efficient 2003. International Waterworks Association. Canary Islands, Spain.

Palau C.V., Arregui F., Palau G., Espert V. (2004). *Velocity Profile Effect on Woltman Water Meters Performance.* FLOMEKO 12th International Conference on Flow Measurement. Guilin, China.

Planells F., Antolí A., López V., Sanz F., García-Serra J. (1987). *Diagnóstico de la gestión óptima de contadores en un sistema de distribución de agua.* Tecnología del agua No 38 1987.

Schlenger D.L. (1997). Meter management: best practices for water utilities. *Journal of Water Engineering and Management*, Marzo; 33–37.

Shercliff (1962). *The Theory of Electromagnetic Flow Measurement.* Cambridge University Press, Cambridge.

Sisco R.J. (1967). The case for meter replacement programs. *Journal of AWWA*, November; 1149–1455.

Sullivan J.P., Speranza E.M. (1992). Proper meter sizing for increased accountability and revenue. *Journal of AWWA*, 84(7); 53–60.

Taborda C., Anunciaçao A.V. (1997). *Os Contadores e a Qualidade da Medição.* Agua Scripta.

Terriel J.C., Daniel W.W. (1994). *Business Statistics for Management and Economics.* Ed. Houghton Mifflin Company. Boston.

Van der Linden M.J. (1998). Implementing a large-meter replacement program. *Journal of AWWA*, August; 50–56.

Yanov D.A., Koch R.N. (1987). *A Modern Residential Flow Demand Study.* Session 27 of AWWA Annual Conference, Kansas City, KS.

Yee M.D. (1999). Economic analysis for replacing residential meters. *Journal of AWWA*, July; 72–77.

Additionally, technical data has also been revised from commercial catalogues of the following manufacturers:

- ABB – Fischer&Porter
- Actaris
- Badger
- Contazara
- Controlotron
- Elster
- Endress-Hauser
- Krohne
- Hersey
- Hydrometer
- Maddalena
- Neptune
- Panametrics
- Rittmeyer
- Rosemount
- Sensus
- Siemens.

Index

IWA Publishing's authorised EU representative for General Product
Safety Regulations is Diane D'Arras, 15 rue Duret, 75116 Paris,
France, e-mail: safety@iwap.co.uk.

Printed and bound by CPI Group (UK) Ltd, Croydon, CR0 4YY
06/07/2026
02157563-0003